Microscopy As A Hobby
A 21st Century Quick Start Guide

Mol Smith
(Founder of the much respected hobby-based
Microscopy web—www.microscopy-uk.org.uk and co-
founder of Micscape Magazine—a Microscopy
magazine of the 20th and 21st century).

ONVIEW BOOKS
Microscopy As A Hobby
A 21st Century Quick Start Guide
Published by Onview.net Ltd
2014

Onview.net Ltd. Registered Office:
Frilford Mead, Kingston Road, Frilford. Abingdon.
Oxfordshire. OX13 5NX England

www.onview.net
www.microscopy-uk.org.uk

The moral right of the author has been asserted.
"Many thanks to Lesley Evans for proof reading the
assembled book and to David Walker for his help and guidance."

First Published 2014 by (Onview Books) Onview.net Ltd.
Printed by www.createspace.com

A CIP catalogue record for this book is available.
ISBN 9781500301651

Microscopy As A Hobby
A 21st Century Quick Start Guide

*A quick start practical microscopy
guide for young people and newbies
in the 21st century.*

Mol Smith

Dedicated to childhood, the prolonging of wonder in those years, and to all mums and dads who share love, insight, curiosity, wonder, and knowledge with their children in a busy world. And written in homage to Bills Ells, John Wells, Eric Marson, Alan Potter, David Walker, Walter Dioni and all the people who contributed to Micscape Magazine and hundreds of people just like me involved with other work aspects but... well... we found this too and helped it survive and grow.

CONTENTS

CONTENTS

CONTENTS

CONTENTS

Introduction. Why Microscopy? Why Me?
(Aimed at parents and newbies).

This book is more suitable for teenagers and adults taking their first steps into the hobby of microscopy. It quickly removes all the confusion and provides an entry level guide into a fascinating area of study which eludes all but the most discerning mind.

You can use this book to learn how to buy a microscope for your older child and then simply hand the book over to them. If your child is very young, you can use this book to help you in getting them started with microscopy as an interest. My work refers to some internet content, most of it mine. This is necessary to include additional resources to improve on what can be communicated about more complex topics in the static media of a real book.

But this work is designed to do something far more exciting than that. It will introduce teenagers and adults to new technologies and methods to enhance and develop hobbyist microscopy and to begin applying it in a brand new era.

Over eighteen years ago as a forty-something year old adult, I looked down a microscope for the first time. By this, I mean a proper one—not those plastic jobs someone buys you as a kiddie for a few quid, and which wrongly informs you that the invisible world at the microscopic level is just a blurred mass of colour. I never saw a microscope in my school days and now in the 21st century, most school children won't see one either. We live in a digital, internet-connected tangle of second-hand experience. Someone, somewhere owns a microscope, takes a video or photo, and posts it on the web. So no one else at school needs to bother looking down a microscope anymore... right?

Wrong!

Alas, so much of the information on the internet is put there by uninformed people and without their posting having any peer review by experienced and knowledgeable people. One path to ignorance is through not learning, the other is by learning false truths thinking them absolute facts. Only when a person takes a microscope, puts something under it to observe, will he or she appreciate the marvel of nature and of life itself. You just can't obtain that experience from a photo or a video.

I have aimed my introduction here at parents wanting their children to have an interest besides computer games and comics. I see nothing wrong with the latter but I see a big advantage in having your young ones interested in science, in life itself - and hopefully using a microscope as part of a hobby will see your children outside in the real world collecting things to look at. You can even join them on nature walks, by the sea, or when on holiday collecting things to bring back and look at more closely. My book is not aimed at the seasoned microscopist. Any adult starting out will find it helpful. It will aid their journey towards becoming an experienced amateur scientist.

We live in an era entirely remote from the Victorian one—the time when amateur microscopy spawned itself as a novel pursuit and as a form of educational entertainment for the wealthier and educated Victorian middle and upper classes. Few things have actually changed in the traditional form of the 'hobby'. And yet we live in a world of HD 3D movies, 3D computer modelling, LED screens, Lasers, Computers, and Internet technologies. My book seeks to address some issues I have with the reluctance of the hobby to step forward in bold and imaginative ways to present itself in the 21st century. My aim is to equip the young novice microscopist as well as the mature enthusiast microscopist with the know-how, tools, ideas, and direction to profit their own pleasure in this pastime, and to help evolve the pursuit further.

There have been films, toys, and games inspired by huge life forms like dinosaurs, which I'm certain have helped capture the imagination of our young generations and inspired them later in life to enter professional areas of research or to improve their knowledge of Natural History. Hardly anything exists in the popular media to achieve the same for the tiniest creatures and plants which dominate our planet. My book therefore is designed to put forward ideas to achieve that into the minds and hearts of all microscopists and for them to see themselves as champions and ambassadors of an area of science, which although was popularised extensively 200 years ago, has since failed to grab the imagination and attention of the masses in ever-declining steps.

We are in fact in danger right now of losing this pastime from human endeavour. That would not only be sad, it would be catastrophic. Nothing can better a human-being with a sharp and curious mind physically taking steps to go 'hands-on' and explore our world and universe. Sure, our technology can go and do it for us, and send us information to our computer screens. I suspect I could build a small drone equipped with a submersible macro or micro lens to drop in my pond and send me back pictures of tiny life forms in there, but if I go myself, I smell the air... feel the sun on my back... and I am the life moving amidst life. I believe that is important. Yes... adapting technology into our pastime can improve our physical and active endeavour, but it should never replace it.

No-one yet can provide an answer to the origin of life. Evolution, a theory, may provide explanation of how living forms adapt and change to their environment over generations, but it offers no cause for the emergence of life in the universe. I think though that we can all assume that whatever the original spark was, and whatever the process is which teases living, thinking, forms from out of inorganic matter, it is accomplishing it by building life bottom up, that is—by making it from tiny structures: from atoms, then molecules, and then cells. Life is not sculptured out of a block of granite. It's built as though from Lego bricks into a final, often complex, whole. A microscope demonstrates that process and therefore gets a young person to realise a great truth immediately. Living things including your child, are created by a code already buried in existing organic matter, so is everything else. Whether that code is put there by God, Nature, or random presentation of a mystical universe, or by Odin himself, is a question of faith or a lack of it. But whatever one's belief, using a microscope can only strengthen all beliefs simultaneously— which is no small thing in a world where different belief systems seem to cause so much discord between groups of human beings.

So here are a few real advantages for helping your child pick up the hobby of microscopy:

1) It costs very little to start.
2) It is a hobby for both winter and summer.
3) It is one of the few remaining sciences where an amateur can contribute with fresh discoveries and insights.
4) It takes children outside and gets them involved with the physical world.
5) You can spend time together with them.
6) It hones curious minds.
7) It can strengthen a love of science and inspire future employment aims.
8) A child can put it away in early teens and take it up again later in life. Many people do.

There are many other good books around to help young people get involved with microscopy. I will provide a small list of those I like and think will help. Because of this, my book is aimed at something quite different. I said earlier that we live in a brave new era of technology, the internet, hi-tech 3D movies, and rapidly emerging new and novel technologies. My book—**and this is important for you to understand**—also integrates microscopy with those other platforms around your growing child. Most books teach microscopy the same way the English Victorians did. I show microscopy as a hobby fit for the 21st century and I try to inspire all who read this to evolve the pursuit of this science still further using the contemporary tools and technology of today.

In the near future, it is a microscope which will look for life on other planets, it is a microscope which is required to help build ever smaller bots and nano-machines; it is a microscope that will be used by future generations not merely to look at micro-forms in the pond, but micro-components in their bodies—our children, as medicine integrates our nano-technologies with our biological ones.

I have created this book in black and white. I have done this to ensure it is inexpensive to purchase. Colour printing is still very expensive. Seeing the microscopic world in colour for real is best.

Now... about me. What qualifies me to create this book? Am I a scientist? No. I guess I think I am qualified because I co-founded Micscape Magazine—a web based magazine for amateur microscopists running since 1995, and I produced the first software program to emulate a microscope on a home computer. I am also an amateur microscopist (we like to say *enthusiast* microscopist by the way as 'amateur' has other connotations). I've been involved with providing many articles, videos, and images for our web presence for what feels like a lifetime, and I am convinced this is the best hobby for any curious and intelligent young person or in fact—for anyone with an enquiring mind.

Mostly I think I am qualified because like many young people growing up, I was looking in the wrong places, and missed something. I missed this and I so dearly wish I had encountered it earlier. Microscopy and the study of the very small continues to fascinate me and has enabled me to look at all of life, and aided my decisions about how to manage it. My endeavours have been more successful through the lesson of looking at the detail of things going on around me instead of accepting the sound-bites, generalities, and noise so prevalent today.

You want your child to succeed in a confusing world where truth is confounded by misdirection and global acceptance of sweeping media statements. Microscopy, as a pursuit, trains the mind to look further, to perceive not just the micro world, but the one we live in day by day. Put simply—it illuminates where ignorance often prevails.

Mol Smith (2014)

Part 1
Entry Into Traditional Hobbyist Microscopy

Chapter 1. What you need to start.

Buying a microscope—which one for the first time?

You need a microscope. Right? Problem is when you look at all the stores on the internet, it gets a little confusing. So many different microscopes, so many companies, so many specs! Just how do you sort out the wood from the trees?

Well, here's a secret. Nearly every microscope sold today is manufactured in just one country. It doesn't matter which company is selling it or the brand name. They are all being made (mass-produced) in China... just like everything else. And as we've all learnt by now, especially if you've ever tried sharpening a Chinese-made pencil, the quality can be...er... shall we say for fairness—variable?

A good company will buy these Chinese produced microscopes, slap about another 100% onto the cost price, and then when you order it, go over it and correct any quality issues before sending it out to you. A bad company will just wrap it and send it, hoping nothing is wrong with it, or knowing if there is something wrong with it, as a newbie— you're not going to notice for ages if ever at all.

I can recommend two good English companies if you are in the UK: Brunel Microscopes Limited, and Apex Microscopes Ltd. I know they are good because I have dealt with them for years and worked alongside them as they supported the millions of viewers of our web site and magazine. They are not the only good UK sellers, just the one I know well. See appendices at the end of this book for contact details.

If you are outside the UK, you should look into the various forums, or on the Amazon web site and at the user reviews there to choose a reliable company. Remember... microscopes are heavy. It is always best to chose a seller in your own country to keep shipping costs low. These can be considerable if the microscope is a large one.

For your very first microscope or for your child's first microscope, this is what to consider. First of all—age. A child below say the age of nine or ten years might find it far more difficult to use some types of microscopes commonly called monocular (or compound monocular) microscopes. There is also a third type called a trinocular microscope., and several others besides. For now, to start the hobby, let's just discuss two.

The monocular microscope

This has a single tube, a bit like a telescope in reverse. It has the capability to magnify objects up to around 1600 times, but only under optimum conditions, and using sophisticated know-how and technique. This is the type of microscope most of us think of when we think of microscopes. It can be used to look at live specimens, both plant and animal, but only if they are very, very small—microscopic in size. There isn't enough room normally to put a bigger thing like a large insect onto the viewing stage because the gap between the lower lens and the stage is not very big. Plant leaves and sewing needles can be put easily into that space though. Look at the images below.

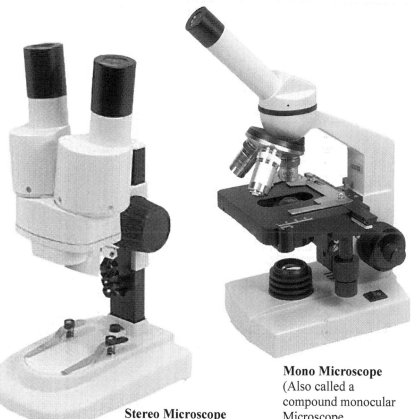

Mono Microscope
(Also called a compound monocular Microscope
One tube.

Stereo Microscope
Two tubes.

The stereo microscope

Stereo: *two.* So, two tubes. Two tubes which magnify the subject on the stage and then combine them into a single image in 3D. You can see depth just like through a pair of binoculars, or at a 3D cinema. Just look how large the gap is between the stage (where the clips are) and the tubes at the bottom of the microscope body. The mono microscope only produces a flat image as you can only use one eye to view anything so your natural stereoscopic sight is disabled. Some stereomicroscopes can magnify up to several hundred times but these can be very expensive due to the two tubes and the increased number of lenses. At affordable levels, they usually magnify 20 or 40 times. And quite honestly, as these stereo microscopes are usually employed to look at insects, gems, and items or living things larger than microscopic in size, 20x or 40x is more than adequate. Another type of microscope is a binocular microscope, which has two eyepieces but only one objective lens. This shares the image to both eyes but not in stereoscopic vision. It is simply flat. Be careful not to be confused on this.

User's age

I think children under the age of nine years old are best given a stereo microscope. They are easy to use. The child can get lots of things to look at from the garden or your home and quickly see them at a huge size without much fiddling around. You merely have to switch the microscope's LED light on, turn a knob on the microscope to focus, and *POW!* It's there, clear as day. The young novice microscopist doesn't have to kill anything either, which I think is nice. He or she can watch eyes moving, legs walking, and antenna shaking in glorious 3D... providing of course they are still attached to the body of the insect.

Above this age, the compound microscope is probably more suitable. It allows more serious study of a truly microscopic world although a bit of knowhow is required for them to get the best out of the instrument. So...

Child aged 8 years or younger: buy them a stereo microscope.
Note: See alternate microscope for youngsters—page 22
Child 9 years or older: buy them a mono microscope.

Buying a stereo microscope for very young children

First and the most important point: *do not buy a microscope from a toy shop.* 'Toys Are Us' means what it says. A microscope is not a toy, which is something you play with and can be made out of all kinds of plastic. Real microscopes have **glass lenses**. Toy ones don't. You can only see macro and microscopic objects sharply and clearly with well made glass lenses. Microscopes in lavish and exciting boxes screaming out exaggerated claims on the shelves of these successful stores have done more to harm the pastime of Hobbyist Microscopy than any other factor. Children get these toy microscopes as presents and believe what they show you is all you can ever see of the microscopic world, which will be nothing but a blurred mess, when it should be a stunning and clear vista of a secret world.

The body and stand of the microscope should be made of metal. You need something that is not going to wobble around or vibrate, so you need metal as it is heavier and sturdier. Your stereo microscope needs a light. Some have an under stage light, some have an over stage light, and some have both. Both is best, but if you can only afford a stereo with one light, choose the one with over stage lighting. This will be called upon far more often and will be suitable for 99% of what a child will look at.

Beam me up, Scottie! Lights need electricity. In days gone by we used tungsten filament bulbs to provide artificial light which also gave out as much heat as light. They also needed a good supply of electricity and didn't last long on batteries. We are now in the 21st century and LEDs (Light Emitting Diodes) provide cool white light efficiently and with only a little demand on batteries. A stereo microscope fitted with an LED light, which runs only on batteries, is a great safe choice for very young people. No mains supply attached, and no transformer, means ultra safety. You can buy stereo microscopes which use lights powered via the mains, but in this instance—you don't need them. You won't need to keep buying stacks of batteries to keep the light working, so long as you teach your young child to turn the light off when they have finished using the microscope. Some stereos have a light level switch or dimmer. You don't really need this for a low cost stereo microscope for a child.

A focus knob winds the stage (or tubes) up and down a rack, further or closer to the subject being viewed on the stage. It's a bit like using a magnifying glass. You turn the knob until what you're looking at is in focus and then you stop. The tubes should not drop down further but stay there exactly where you stopped them. If the focusing mechanism is sloppily made, the tubes will drop further and focus is lost. This means you have to wind it up a bit and keep guessing where to let it go. This will be very confusing for a young child.

The focus might be enabled by two knobs. One is called coarse focus, and the other is fine focus. This is a more expensive feature of a stereo microscope when compared to a single focus one. The idea is that you roughly focus with the coarse focus and then refine the focus with the fine focus control, which moves a tinier distance for any part turn of the knob. It also allows absolute focussing on any part of the subject being studied. Most children will be okay with just the single focus control, providing it is well made and not sloppy.

No two eyes are the same. Some of us have eye defects—short-sightedness, long-sightedness, stigmatisms, or just plain old slight differences in the focussing efficiency of left eye to right eye. To get over this when looking through a stereo microscope (and binoculars), one of the eyepieces can have a small adjuster on it, and sometimes both are fitted with them. They are called eyepiece Dioptric adjustment rings. Your stereo microscope must have at least one Dioptric adjustment ring otherwise children and adults with less than perfect stereo vision (most of us) are not going to achieve clear and sharp viewing.

The space between our eyes differ from person to person. It's called the interpupillary distance. Your stereo microscope must have tubes which move wider or closer together to allow for comfortable viewing for people with different distances between their eyes.

Low cost basic stereo microscope

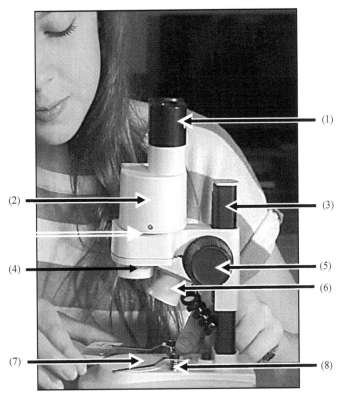

See the image above and my summary notes for the main points on buying a stereo microscope for a younger child.

1)   Eyepieces. Glass! At least 1 to have Dioptric adjustment.
2)   Both tubes adjustable for differently spaced eyes.

3) No sloppy rack and pinion focusing.
4) Coated **achromatic glass** objective lenses.
5) Stable focus control. No dropping after focus set.
6) LED battery-powered over-stage light.
7) All metal body and tubes!
8) Slide clips—useful but not a necessity.

The objective lens, down close to the stage can vary but on low cost microscopes are normally fixed (you can't swap them out). In the one shown on the opposite page, the objectives are 2x. This means the total magnification for this microscope will be 20x with the standard eyepieces of 10x.

Stereo microscope prices
The budget one shown costs, at the time of writing, around £50.00. If you find microscopes for sale on Amazon, remember that Amazon add their profit slice to the price. This is included in the price you see but it's worth tracking down the company selling the microscope and finding out if you can buy directly from them. This way, you won't have to pay the extra money added by Amazon.

Many companies now selling microscopes on Amazon may have several company names: one to sell through Amazon and one to sell to you directly. Guess which one will sell it to you cheaper! Carry out a web engine search for the company name and you'll easily track down their real web site instead of their Amazon store presence.

Check with the company what the next model up costs and consider the advantages of that model and the extra amount you would pay. Often the one up from the budget model is better value in respect of the extras offered. The cheaper or so called 'entry' models are almost always good, but often may lack that one thing you would want once you get a taste for the hobby of microscopy. All the companies selling microscopes are trying to compete with others to seize your attention as a first buyer, so they have to strip back all but the essentials—yet the next step up in a microscope model will include the little things you will wish you got for your child if they end up loving the hobby. For a first stereo microscope which is good value and robust, with good quality optics, excellent durability and hassle-free, you should budget for between £45 to £85 in the UK plus shipping costs. It's cheaper still in dollars if you live in the USA as many items sold in Europe and England are cheaper in the States than in Europe. If you live in the European community, translate the UK sterling value to Euros and look for the equivalent in your country. I write this in 2014. Inflation over time works out at around 3% per year—if you wish to equate these prices to a later era.

Tweezers, dishes, glass slides, insect catchers
Forget all this for the younger child and the stereo microscope option. Most of the time, what you have already in your home will be ample. But if you buy the compound microscope, *buy some blank glass slides too, some cover slips, and an eye-dropper.* The older microscopist will need these.

Cameras
Yes. Cheap digital cameras are available to put in where the eyepiece goes but an eight year old or younger will get more pleasure and a better insight by using a simple sheet of paper and a pencil and drawing what she or he sees looking directly down a microscope. When you draw, you remember and develop.

Buying a monocular microscope for an older child or newbie adult
Some of the things I discussed in buying a stereo microscope for a younger child also apply here. But in case you skipped that part, I will take you through these aspects completely again with the risk that you did read the previous section, and the repetition will bore you. Just skip the bits you already read.

Metal and Glass
If you ever tried to use a telephoto lens on a camera or held a pair of very powerful binoculars, you'll realise how important stability is to obtaining clear, sharp, still viewing. Even the tiny imperceptible quiver of a blood pulse in your vein will move your hand and shake the instrument. A microscope is more powerful than both binoculars and telephoto lenses by orders of magnitudes ten or twenty times. This means any tiny vibration—a train passing by on the rail system 100 metres away—can shake a microscope on a desk simply through ground vibration. You need cast metal for the compound microscope or strong alloy so the instrument is weighty enough to sit still and strong enough to withstand wear and tear over years. Nylon and plastic parts make for cheap manufacturing costs but they also make microscopes which are mostly unusable for any serious observation.

Microscopes work by bending and focusing light waves reflected from, or passing through transparent specimens. Reshaping bundles of light waves into a coherent cone of light up through a microscope tube is fraught with problems solved only by expert manufacturing of precisely shaped glass lenses. Often, several elements are required in a lens system to manage aberration and colour scattering, A very complex area of science and mathematics is applied to design the right types of lenses for the job. Only one material can make perfect or near perfect lenses. It is glass. Nothing else will do the job well. Your monocular microscope must have all glass lenses. Good glass objective lenses are also coated to help eliminate reflection/refraction problems. *Always ensure objective lenses are coated achromatic glass objectives!*

Light

The more you magnify something, the dimmer it gets. You need a good light below the stage as almost everything you look at under a monocular microscope will be illuminated from beneath with the light passing through the subject (always transparent or semi-transparent) up into the lens and tube system. An LED light is now the best light because it's bright, white, and cool. Tungsten bulbs, as lighting systems used in microscopes made before the 21st century, produce a lot of heat which is not required. It's best to use a mains powered lighting system to power the light on a compound microscope as the LED is more powerful than on a low cost stereo microscope and therefore demands more power.

The light need to be dimmable. Different subjects being viewed will have different grades of transparency. Also, different magnifications require different intensities of light such that the more you step up the power and magnitude of magnification, the more light you need to see anything clearly. Your microscope light should at least have a switch which selects several levels of brightness. In an optimum system the light will be controlled by a dimming knob.

Light from the base of the microscope is further managed by passing it through a condenser to concentrate the light in a loosely shaped beam which then passes through an aperture to eradicate spurious light around the fringe of the beam. Apertures are also used on cameras. Most are iris apertures, which is one where a series of overlapping thin blades open and close to form an adjustable hole at the centre. Closing the aperture reduces the hole. Opening the aperture widens the hole. Low cost microscopes may not have an iris aperture but instead have a rotating plastic disc contain three or four different sized holes. This makes for cheaper costs and yet again also makes viewing options less than perfect. Always ensure a compound microscope has an iris aperture below the stage. More expensive microscopes have another iris aperture as part of the condenser arrangement.

In addition to the light, condenser, and iris aperture, compound microscopes can have several 'extras' to aid the management of light through the specimen up the tube and to the eyepiece. These can include filter holders, filters, and a polariser which can be very useful when viewing certain objects. I think though the extra cost involved for anyone trying microscopy for the first time makes these extras unnecessary. It is best to purchase a low cost good quality compound microscope which has the basic parts and which works well, and see how involved the budding new microscopist becomes in the pastime. Once someone has been practising microscopy for a year or more, they can decide how deeply they wish to get involved, and that will determine the level of sophistication they require for the acquisition of a new and more expensive microscope.

A compound microscope may also have an over-stage light but in my opinion, this will not be necessary for early and first studies and is only rarely used at all by seasoned microscopists. Most subjects studied with a compound microscope are best viewed by light passing through them from beneath.

Lenses and magnification
The total magnification of a compound microscope is derived by multiplying together the magnification factors of the eyepiece lens and the objective lens. On all but the most rudimentary microscopes, different types and different eyepieces and objectives can be changed out for others. One might have, for example, x10 and x15 eyepieces. Normally they are changed by just gently pulling out the top lens and popping the other one in—a slide-in fit!

The objective lenses are supporting in a revolving turret at the bottom of the turret, just above the stage. There may be three or four lenses in the turret. A typical arrangement is a set of objectives with x4 x10, x40, x100 (oil) magnification. You might note the 'oil' tag on the x100 lens? This indicates the lens is designed to dip into a special oil droplet placed on the coverslip of the observed slide. The oil removes the air gap and improves the path of light coming up from the slide through the lens into the tube. At high powers of magnification air, normally present in the gap between the slide and the lens, bends the light too much and therefore degrades the clarity of the information transmitted in the light cone. The special oil corrects this problem enabling a sharper clearer image to be seen.

A microscope supplied with lenses as specified above will have a magnification range of x40 to x1600. The highest level of x1600 magnification is close to the limit at which a compound light microscope can magnify. The limit is imposed by the wavelength property of light and the limitation of light waves themselves to transmit more information. Take note that most microscopists using a microscope as a hobby will rarely need to magnify specimens more than 500 to 600 times.

If the microscope you are buying for the first time does not have a x100 oil immersion objective lens, or the microscope cannot magnify more than say—400x, I shouldn't worry too much as most young people will find this more than adequate unless they really get involved deeply into the pursuit. Also note that bacteria are extremely difficult to see more than as a few dots or dashes under anything but very high power, top spec, top price microscopes. And they are inevitably boring subjects to look at for all but the research scientist. Microscope suppliers might try to sell microscopes based on claims of 1000x or more magnification but you should ignore this and opt for a good mix of lenses which provide the greatest variety of magnifications between 50x to 600x: a far more useful choice.

Mechanical stage
A full mechanical stage with drop down coaxial controls is really a must if the microscopist wishes to look at living things which are very small—the sort of things you find in ponds. So, what is a mechanical stage? It moves. More than that it moves very smoothly along both x and y axes which means simply left and right, north and south.

Two controls... small knobs below the stage, allow the microscopist to move the stage very finely in very tiny movements such that anything swimming around in a tiny drop of water on the slide can be kept centred and in view. The mechanical stage also allows searching the slide for interesting processes or life forms to look at. Without this fine movement through mechanical control, the microscopist must user his fingers to nudge the slide around. This resembles huge left /right jumps to the observer and will frustrate anyone who wants to look at something small in detail. Always try to afford a compound microscope fitted with a mechanical stage. Stage clips are almost always supplied to both fixed and mechanical stages. These keep a slide steady and flat. See image and summary below.

The price of a very good basic level monocular microscope with a mechanical stage and iris aperture at the time of writing is approximately £150 to £200 (British Sterling Pounds), plus VAT. Cheaper in the USA.

This might seem an expensive gift if one is buying it for a teenager but it is very good value compared to the price for game consoles. The microscope can always be sold on later at a good second hand price on EBay if the young microscopist decides to move on to a better microscope or loses interest.

Testing what you buy

As I mentioned earlier, almost all microscopes are made in China and hopefully, whatever company you bought yours from did a quality check before sending it out. But... just in case... best to check it out first. The simple bits to check can be done first.

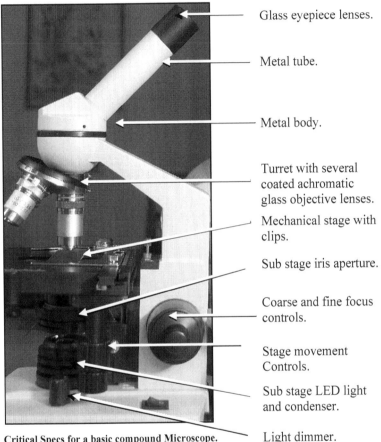

Glass eyepiece lenses.

Metal tube.

Metal body.

Turret with several coated achromatic glass objective lenses.

Mechanical stage with clips.

Sub stage iris aperture.

Coarse and fine focus controls.

Stage movement Controls.

Sub stage LED light and condenser.

Critical Specs for a basic compound Microscope. Light dimmer.

Check list 1.
- Are the eyepieces clean and not scratched?
- Are all the parts there, including any additional lenses?
- Any screws missing or wobbly bits on the microscope?
- Set it up and test the light works, and the dimming control.

All good? If not, best to contact the supplier right away to see if they can correct whatever is missing or broken. If all is well, do the finer checks:

Check 2

The centred light cone/train requires all the lenses, the light, the sub-stage condenser, the iris aperture, the objective lens, and the eyepiece to be perfectly centred. The sub-stage condenser and the iris aperture might (probably will have) small adjustment screws to help alignment but don't fiddle with them, unless there are deliberate controls for that. Screws requiring Allen keys or tint screwdrivers should be let alone.

Place a specimen on the stage, turn on the light and dim to low, then focus under a low power objective (10x or 5x). If testing a stereo make sure you adjust the tubes' interpupillary separation, until you see a single circle. Focus with the coarse control slowly until you can see the object, then refine your focus using the fine focus control [if it has one] until the object is sharp and clear. Be careful not to wind the slide up into the objective lens!

The aperture diaphragm (also called an iris diaphragm) controls contrast, and is found in the condenser, which sits right below the stage in line with the microscope objectives. The condenser may be movable, both in the horizontal and vertical directions. If the condenser is fixed and has no position adjustment, it has been pre-centred at the factory, but it should still have an aperture diaphragm with a movable collar or knob. With a bit of practice, you can make adjustments by viewing the image quality. The aperture should be closed down approximately 1/2 to 1/3 for proper use. A practical method of adjustment is to open the diaphragm completely and then slowly close it down while viewing the specimen. As soon as you see the contrast improve, leave the diaphragm at that setting. This microscope aperture diaphragm adjustment is critical to achieve the optimal image of the sample. The collar or slider adjusts the degree the aperture is open or closed, thereby affecting the depth of field and overall image quality. The aim is to achieve best resolution with good (not too much) image contrast. .

Sometimes, centering is not precise on low budget scopes and you just have to accept that.

Check 3

How good did the focus work? On low cost stereo microscopes with only one focus control, sometimes its not always possible to achieve an optimum focus, but when a microscope has fine and coarse focus, it should be easy to set the focus and not have the microscope slowly go out of focus when you let go of the focusing controls. If it does, contact the supplier and tell him/her the focusing is sloppy and return the microscope. As an aside, a lot of well used second hand professional microscopes, costing far more than entry level hobby microscopes, end up with sloppy focusing through worn rack and pinion parts due to years of constant use. They can sometimes be corrected by using packing grease, but this is not something you do on a brand new microscope.

At very low prices, you cannot expect to obtain a microscope with the fine engineering and precision of instruments costing thousands of pounds. So, please keep that in mind when checking the quality and centering of the microscope. It may not be perfect but it should be good enough to make the instrument a practical tool.

I forgot to mention about the Dioptric adjustment on one or more of the eyepieces of a stereo microscope. Always focus first with one eye closed and looking down the eyepiece without the adjuster first. When you are focused there, open your other eye and match focus by using the dioptric adjuster. You can always open one eye, close, it, open the other, back and forth until you are satisfied focus is good for both eyes.

Important: an alternative microscope for very young children
Some people have issues looking through binoculars and stereo microscopes, often when they do it for the first time. What happens is the two circles do not combine into one circle. This happened to me once after not using a stereo microscope for some years. I looked down it and thought something was broken because try as I might, I could not get the two circles to combine. It took me a few minutes before it dawned on me it was my mind itself refusing to combine the images. As soon as I realised it,

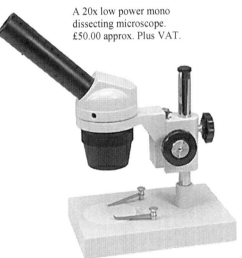

A 20x low power mono dissecting microscope.
£50.00 approx. Plus VAT.

magically—the two circles came together. See: it's a trick of the mind, but remember this if your youngster struggles with the same. Often 'thinking the circles together' coaxes the obstinate mind to conform.

An alternative microscope for very, very young children

Some very young children struggle with stereo microscopes. It may be their interpupillary distance is shorter than an adult's (distance between eyes) or their minds just cannot combine the two images into one.

There is a solution but not one I necessarily recommend. You could buy a very young child a low powered monocular microscope. Although they are not specifically designed for children per se, they are easy to use and quite low priced. But they come often with no frills like lights. But you can always use a cheap LED torch to illuminate a subject under study. I think a stereo is best because if the child can use it, they will discover the 3D experience provides more

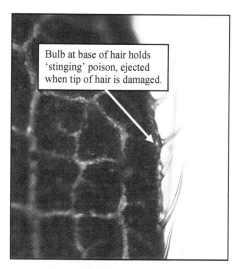

Bulb at base of hair holds 'stinging' poison, ejected when tip of hair is damaged.

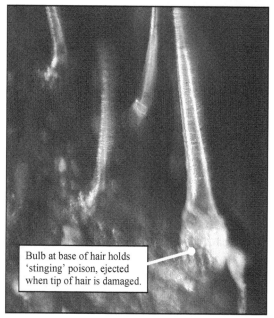

Bulb at base of hair holds 'stinging' poison, ejected when tip of hair is damaged.

detail and is a lot more fun. A typical low power monocular like the one shown from Brunel Microscopes will cost about £50.00 in the UK.

You should ensure the build quality is at the same level as described for the other microscopes detailed in this chapter.

So... just as a quick guide to the capability of a low cost stereo compared to low cost compound monocular microscope, I quickly photographed a nettle using exactly these types of microscope. The one [top, previous page] shows nettle stinging hairs at about 20x magnification with the stereo microscope. The one below it [previous page] shows the same nettle (different hairs) at a mid-level magnification using the monocular compound microscope. *Note: not the dissecting low power monotype.*

I've cropped from the original photos which show more of the leaf and are huge colour photographs. A simple adaptor costing just a few pounds allow most digital cameras to attach to the eyepiece end of the microscope (or a video camera) to record things studied.

Summary
There you have it. I would recommend when you buy your first microscope, you also buy the books I recommend in chapter 2 and keep them close to my one. I see little point or value in repeating information contained in these other works, (I will, a little) but what I have to say in my work will help pull together what you learn and move you forward through some of the things they get wrong—often because... well, the world has changed since these books were first written.

One good example is that many of the chemicals used in microscopy to preserve and mount specimens onto slides are now deemed toxic, dangerous, or just simply banned. I will help you solve that problem because there are things you can obtain simply, from food stores or local shops, which you can use to do exactly what is required to make long-lasting specimen slides.

Another example is that few existing books create a vision of the future or challenge new microscopists to move forward through the 21st century to discover fresh applications for microscopes and the science of microscopy. I will help show you the way and guide you into totally new and novel directions for microscopical application. Here in is the key to success for future pioneers of hobby microscopy!

Promoting the hobby and the science
You may be thinking of buying for a youngster or buying for yourself. In both cases, some things are very true. In a world where we increasingly seem to immerse ourselves in work, and where children and adults too are gradually being embedded into a technical and prescribed environment of chasing a better life experience through qualification, more work, and less leisure time—an escape into the random freedom displayed in the natural world can be a great release.

I have, as an individual, lived a corporate existence for a top technology company, been an engineer, a tradesman, an author, a film-maker, and I've journeyed through many invocations of what a person may become or experience. But I can tell you right now, nothing can compare with an afternoon off, a walk in the country, by a river, or sitting by my garden pond watching dragonflies and damsel flies come and dance before my gaze and delight me with their performance.

And taking something back from a walk or dipping into my pond and placing a drop of water under my microscope, or a piece of a new plant or a soil example, has allowed me to understand not only more about the world I live in, but the larger world of people, systems, organisation, construction, networking, parasites, deception, truth, and purpose. I don't think I'm alone. The 'wow' of a young person seeing something new and often repulsively different, can in a short time become a nod of a head of a mind within, exploring and considering silently far more profound things, as he or she looks into that revealing beam of light. Of course there are lots of quick hit alternatives, micro

www.pippasprogress.net

Pippa's Progress. Videos for children on-line.

cameras for Iphones, Youtube videos (most done badly and boringly), but the only way to really know a thing properly is by doing it well.

You may be an adult looking for something else in your life to add meaning and substance to it besides the daily drudge, and family, work, holidays away, and movies.

I suggest you give it a go. Buy your first microscope. Read the rest of this book, visit our web site at : www.microscopy-uk.org.uk and discover not only what fascinated our Victorian forebears, but has become the domain of silent intellectual people all around the world.

Chapter 2: Learning how to practise microscopy

Buying a first microscope, especially for a child or young teenager, is only half useful. Unlike, say football, or swimming as a pastime, microscopy requires resources, self-learning, and a bit of training. I mean if you want a daughter to do ballet dancing it's no good just buying her pink tights, a tutu, and ballet shoes, then expecting her to go off and start ballet dancing on stage. She needs to learn and she will probably go to a dance school to learn. Microscopy at its basic level, and the way most young children encounter it up to now, is about putting things under a microscope, looking at them, and discovering objects, plants, and animals normally invisible or too small to see well with the naked eye.

There are no real microscopy schools for young people and training of any kind is normally reserved for adults. I know that Brunel Microscopes in the UK provides training and tutorial sessions for hobbyist and professional microscopists but they are not really in the business of running training courses for youngsters. There are various hobbyist microscopy clubs in most of the western countries. Traditionally, they seem to have very few junior members.

The way children are going to get assistance is via the internet or through books. And there are absolutely hundreds of them. But many are just too advanced for young people or adults starting out. Most of the books for children replicate processes and ideas which have been around sine the Victorian era. In one of them for example, a wooden cotton reel is used as part of a method for making thin slices of plant stems to look at under the microscope. Er... you try and find a wooden cotton reel today; impossible in my town. They are now made of soft plastic with a skeleton face, which cannot be used as the wooden ones could as a suitable flat cutting surface. Many books refer to chemicals and stains no longer available to adults, let alone minors, because our over-safe-centric societies deem them toxic or too dangerous.

My web site at **www.microscopy-uk.org.uk** and our online magazine at **www.micscape.net** has thousands of articles and images regarding modern techniques to help hobbyist microscopists, including young people,. There are many other web sites which I will put into the appendix of this book. Some are so important that I will list a few of them here *(see next page)*. I have also included a few must-buy books which I believe will supplement this book and makes it unnecessary for me to repeat what other award winning works can teach the young or newbie adult microscopist. As I mention in my introduction, my book is not about teaching microscopic techniques per se: it is about getting a new person or child up and running quickly practising microscopy based on the past, but now extended into a new age due to changes in technology and know-how. There are books which appear to be about microscopy, but in fact are not: they are collections of wow-factor pictures normally taken by a super-powerful electron scanning microscope. These will not help anyone learn how to use a microscope or give them any real insight into what they can discover in their optical microscopy explorations.

To keep your initial costs down, I will recommend these two books only as perfect first books for children and adult newbies. I believe you should buy them straight away and consider these two and

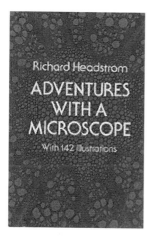

Note: this book is in full colour

Adventures With A Microscope
Item model number: FBA-|298859
ASIN: B0006D6PSM
Price: £10.00 approx.

The World Of The Microscope
Paperback: 48 pages
Publisher: Usborne Books
(1 Jan 2008)
Language: English
ISBN-10: 079451524X
ISBN-13: 978-0794515249
Price: £6.00 approx

TIP!
If Amazon are out of stock, try Brunel micro-scopes in the UK

my book will form a complete school for young and apprentice hobby microscopists to take them from novice to the next step. I will be referring to them in some of my chapters as even these two wonderful works still have a few legacy items from the past which need modifying for today's world.

Remember that microscopy spans a multitude of study areas. Just as there are different dance styles, where a trainee dancer might try several styles before specialising in one, microscopy has topic areas: pond life, parasites, ocean life, insects, plant life (botany), biology (human and animal), and many others. Books exist which specialise and go deeper into these sub-areas of study. I suggest you don't get involved with them yet, let's get up and running first with the general study and see what might evolve from there.

Pond life

There is no doubt that the most engaging area for young microscopists is studying what lives in ponds, rivers, and oceans—also in gutters, water lying on manhole covers, bird baths, puddles, gold fish bowls aquariums, etc. There are thousands of unique swimming, crawling, spinning, flapping, diving, microscopic plants and animals capable of predation, snaring, catching, and colour-changing behaviours. And most are easily studied alive by simply putting a drop of water on a slide under the microscope.

Curious people are often intelligent people too. They will want to know the names of the creatures and micro-plants they see. Well, even experts often find it difficult to identify any given organism, so it's best if they learn the major ones which are readily recognisable when they start out. The *bible* of pond life, which is an enormous work was written a long, long time ago, but still remains a much referred-to work when trying to identify entities found in freshwater.

It is over a thousand pages—a mammoth and vital work, unsurpassed since it was written.

Fortunately it is now out of copyright and is in the public domain. The entire work can be downloaded on line, by visiting:
www.microscopy-uk.org.uk/mag/artnov13/ms-whipple.html

FRESH-WATER
BIOLOGY

BY

HENRY BALDWIN WARD
Emeritus Professor of Zoology in the University of Illinois, Special Investigator for the United States Bureau of Fisheries, Etc.

AND

The Late GEORGE CHANDLER WHIPPLE
Formerly Professor of Sanitary Engineering in Harvard University and the Massachusetts Institute of Technology

WITH THE COLLABORATION OF A STAFF OF SPECIALISTS

NEW YORK
JOHN WILEY & SONS, Inc.
London: CHAPMAN & HALL, Limited

It may prove far too dense for most young children but teenagers and adult newbies can search the thousands of drawings to identify many of the microscopic things they discover in freshwater. Start saving and washing all those empty jam and peanut butter jars. They'll come in handy for collecting pond and river water.

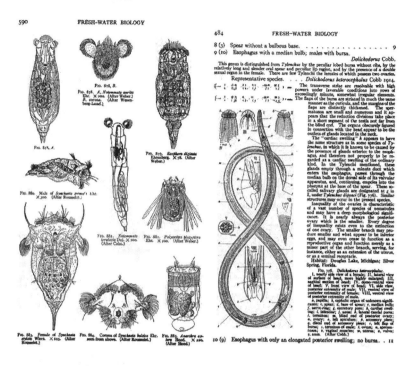

A list of a few great web sites for hobbyist microscopists

Cells Alive. Great for children and adults — www.cellsalive.com

Molecular Expressions. Comprehensive. — micro.magnet.fsu.edu

Micropolitan Museum. Kids love it. — www.microscopy-uk.org.uk/micropolitan/x_index.html

Microscopy-uk. The best (er..it's mine). — www.microscopy-uk.org.uk

Micscape. Vast resources & lists of clubs & on-line forums — www.micscape.net

Brunel Microscopes. Trusted shop. — www.brunelmicroscopes.co.uk

Primer to start with a microscope — www.microscopy-uk.org.uk/primer/

Chapter 3. How to use your new microscope

Once you've unpacked your new microscope and checked it's all there and in good working order, you will be wanting to use it straight away. If you have a stereo microscope, it's very easy to get going. I don't think you need much help with it. Just switch on the light and put something interesting under it to see. You might just want to read the last paragraph of page 13 and all of page 14, paying extra attention to moving the two tubes wider apart or closer together for your eyes, and setting up the focus for each eye. But if you have a monocular microscope you will probably need a little help from this chapter to guide you through getting the best out of it. I'll assume you have never used one before.

First, study the picture on page 19 to familiarise yourself with all the parts and controls. If you never used a microscope before, here is the order in which we set it up to look at something on the stage.

Proper viewing position *(relevant to stereo microscopes too)*
Put the microscope on a solid table or desk. Ensure the height of the desk is convenient for you to look down the eyepiece when sitting on a chair at the desk. If you have to strain to look down it, find some cushions to put on the chair to raise you up higher.

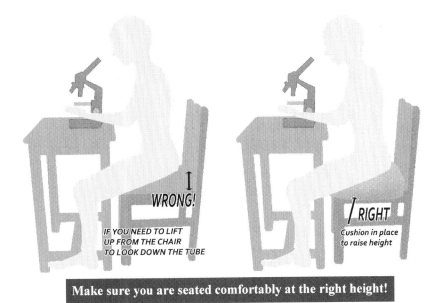

WRONG!
IF YOU NEED TO LIFT UP FROM THE CHAIR TO LOOK DOWN THE TUBE

RIGHT
Cushion in place to raise height

Make sure you are seated comfortably at the right height!

Turn on the light *(relevant to stereo microscopes too)*
If your microscope's light is powered by the mains (plugged into an electric socket in the wall), make sure the lead/cable is not overstretched or running to the socket in such a way you could trip over it or snag it when working with your microscope. You may need an extension lead to ensure safe working and arrangement of the lead. Turn on the microscope under-stage light and turn the dimmer down to low. *You don't want to look down the microscope at a full-on light with nothing on the stage as it will be too bright for your eyes!*

Select a low power objective
Without looking down the tube, instead watch the mechanical stage and the objective lens, use the coarse focus control to wind down the stage to its lowest position. You can now safely turn the turret housing the objective lenses to select the one with the lowest power of magnification, normally a 5x or 10x. Always wind down the stage and look at what you are doing whenever selecting a different power lens. If you don't, it is very easy to try and swing a higher powered objective into place, and discover—*when you crack the glass slide*— there was an insufficient gap between the stage and the lens.

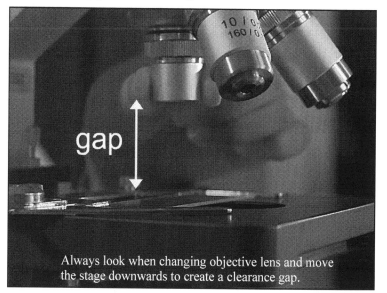

Always look when changing objective lens and move the stage downwards to create a clearance gap.

A sprinkle of sugar and salt *(relevant to stereo microscopes too)*
Take a glass slide and carefully put a sprinkle of sugar and a sprinkle of table salt on it. We will use these as out first specimens. Carefully

slide the glass slide under the stage clips and look down the tube through the eyepiece.

Coarse focus *(relevant to stereo microscopes too)*
Slowly turn the coarse focus knob to bring the stage up towards the objective lens. Make sure to watch carefully and when the sugar and salt crystals come into view... stop!

Fine focus
You can now use the fine focus knob to bring the crystals into sharp focus. It's always worth checking at this point how big the gap is between the slide and the tip of the objective lens. On low powers the gap is much bigger than when using the higher powered objectives. Remember, when using the higher power objectives, it is all too easy to accidently wind the slide into the lens with the focus control and possibly crack the slide or damage the lens. I have done this a number of times myself! So always check this gap when working.

Improving the view with the iris aperture
A lot of information exists on the internet and in various books about the correct use of the iris aperture to improve the detail of the item studied. Which is something that truly amazes me. I think this is a quick and easy method. Wind it up closer or down further from the stage. Open it slightly, close it more. Adjust the under-stage light. See! The iris aperture can change the contrast of the specimen, or wipe it out with too much light, or create an uneven illumination where the centre is over bright and the outer edge is dark.

The best setting for the iris aperture is the point at which you are satisfied with the clarity and detail in the subject being studied.

Over/under stage lighting *(relevant to stereo microscopes too)*
If you have a light over the stage pointing down at the slide, try turning that one on and the under stage lighting off. Do the crystals show up better with that form of lighting? If you do not have over-stage lighting on your microscope, shine a bright torch onto the slide instead. You can put some dark card or black paper under the glass slide when using over lighting to help improve the outlines of the crystals. Try a mix of under and over-stage lighting. Can you work out which crystals are sugar and which are salt?

Different magnifications
Some microscopes allow you to switch out the objective lens to a higher magnification without the need to refocus. Lower priced

microscopes are often claimed by the retailer to be able to do this too, but in my experience—most don't. You will probably have to re-focus again. Remember what I said about looking at the objective lens and the slide on the stage when turning the turret to make sure you have enough of a gap for the objective lens to move into.

Sugar and Salt crystals

Searching the slide (*relevant to stereo microscopes too*)
If your microscope has a mechanical stage, you can use the two controls to move the stage and thus the viewing area of the slide. Microscopists using instruments without a moving stage will have to use a pencil, lollipop stick, tweezers, or something similar to gently move the slide on the stage. This moving the slide, both with a mechanical stage, or manually, takes some getting used to as when you look down the microscope, the movement is opposite to what you expect. This is because the microscope, much like a mirror—inverts the image. Left becomes right and vice versa.

Care of your microscope
The most important thing is keeping your microscope clean and dust free. They are normally supplied with a dust cover which you should get into the habit of using to cover the microscope when not in use. Lenses can be cleaned using special tissues, air puffers, brushes, and cloths used by photographers to clean their camera lenses.

Safety advice

The most likely risk of any accident is when using glass cover slips and glass slides. Cover slips are very fine slivers of glass used to put over a specimen on the glass slide. Because they are extremely thin and fragile, they are easily broken with the chance that tiny slivers of glass can become embedded in your hands. Use tweezers to handle cover slips to reduce the risk of this.

Do not go to outside bodies of water on your own. Should you slip into a river or large pond, and no one is there to help, you can too easily get into difficulty and drown.

When disposing of cover slips or glass slides, wrap them inside folded paper before putting them into rubbish bags or bins.

The second main safety issue is when you collect samples from ponds or streams. Rats may spread a disease into those waters which can cause serious illness and death. It's called Weil's disease (pronounced as though the 'w' were a 'v'). Only a very small number of people die from this in the UK each year, around three, and taking a few simple precautions will protect you: wear washing up gloves when collecting samples, always take a bottle of clean water and a roll of kitchen towel with you to wash your hands after putting them into water outside. The disease is spread through tiny cuts on your skin or via contact with eyes, nose, and mouth. *Note: try to use a microscope with both eyes open to avoid eye strain.*

Things for very young children to look at

Older children can get involved using a compound microscope to study a multitude of things. No so with very young children. For them, you need to select things easy to handle and you'd normally need to assist them. Assuming you took my advice and they are using a low powered stereo or mono dissecting microscope, here is a list of things you can try and what to direct their attention on.

Household, personal, and garden objects

Bristles on a toothbrush (good to educate about cleaning teeth).
The point of a sewing needle (how blunt it looks).
The eye of a sewing needle (unthreaded and threaded. Point out how ragged the cotton is passing through the eye).
A ball point pen (the inky ball at the tip).
Hairs and fibres from carpets (different colours, different thicknesses).
Postage stamps and photographs (colour made up of tiny dots).
Sugar and salt crystal (different sizes and regular shapes).
Ground spices.
Honey and jam (pollen grains and tiny seeds in the mix).

A matchstick, cocktail stick (grain in the wood).
Small coins (the lettering around the edge).
Dead insects (often found around the house).
Rice crispies and other cereals (sugar on surface and different surface textures).
Rice grains—cooked / uncooked (size).
Bread crumbs (texture, air bubbles in bread).
Hair from child's head (colour, thickness; compare hairs).
Finger tip (pattern of lines in skin; compare).
Fibres from clothes, towels, cloths (all different).
Daisy, buttercup, and small garden flowers (petal shape, pollen).
Pollen from larger flowers.
Live or dead garden insects (best not to kill anything; ants are great)
Tiny stones (surface texture, colour mix).
Leaves (the veins, hairs, colour).
Stinging nettle leaves (stinging hairs).
Feathers (fine barbs interlocking the hairs together).
Bread Mould
That should get you started and nothing needs any real preparation.

Specimen slides for older children and adults
You can purchase sets of slides containing samples from plants & tissues. You can also buy slides with mounted insects, pond life, bacteria etc., These are more suitable for young teenagers or indeed—adult newbies. The quality of these can vary greatly from useless to brilliant, which is often reflected in the price.

Below. A photograph taken through a low powered compound microscope of a prepared specimen slide.

Leg of a bee (drone).

39

Teenagers studying biology will benefit greatly from sets of slides relevant to the topics in their examinations: cell division always being one of them. Mounted and prepared specimen sides can be expensive but when well made, will last many, many years. A google search will quickly reveal lots of places you can purchase these slides online.

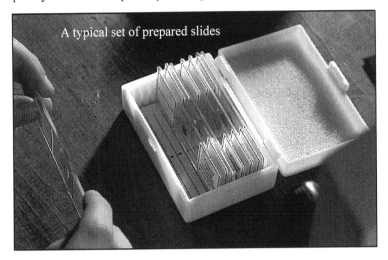
A typical set of prepared slides

Digital cameras

A whole area of microscopy is photo-microscopy and the recording of high quality images and videos. Cheap adaptors are available to enable most high quality digital cameras to attach to the eyepiece. SLRs (single lens reflex cameras) are perhaps the most suitable. There can be issues, not least because you are then using the microscope much like a projector but in this case 'throwing' the image onto the camera's ccd (charge-coupled device—the sensor). Young teenagers and mature adults make full use of the internet with many videos recorded at a

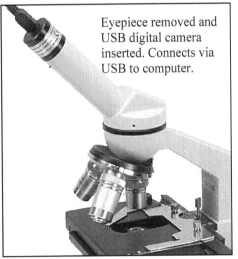
Eyepiece removed and USB digital camera inserted. Connects via USB to computer.

microscope uploaded onto youtube.

A camera can be connected straight to a laptop or desktop computer and the screen used by more than one person to see something on the microscope at the same time.

For young people, a simple USB digital camera (low to medium quality resolution) is affordable and is fitted by simply removed the eyepiece and pushing the camera into the top of the tube (see example opposite). These can be purchased from many microscope suppliers with prices varying. The one shown is from Apex Microscopes and currently costs approx. £50.00. It's a 2 Megapixel camera.

For the older, mature adult taking first steps into hobby microscopy, the quality of these cameras may not be sufficient for any serious recording of videos or stills. It is better to use adaptors to mount your existing SLR to the microscope so you can take full advantage of the higher sensor size, and greater pixel number arrays in modern digital consumer cameras. If you are interested in this aspect, please refer to my chapters dedicated to video and still photography.

All-in-one USB microscopes
There are now completely digital microscopes like the Celestron Amoeba below. You cannot use these like a true microscope: you cannot look down the eyepiece. They connect to a computer and the viewing is always via the screen. I do not consider these entirely useful,

but more a quick fix to show kids 'wow' type pictures of microscopic forms. The cameras are less than 2 megapixels (in this model) and beyond 'the quick look' and ease of use, I consider young people using them cannot directly progress to more serious study. I do, however acknowledge their use in infant classroom and for very young infants at home. They may help promote interest early and thus encourage a child to desire and obtain a real microscope when he or she is a bit older.

The price of this type of USB microscope in 2014 is approximately £60.00 – £80.00. The camera microscope can be removed from the stand and used less statically.

Chapter 4. Basic first projects for microscopists

Earlier, I recommended you also buy *The World Of The Microscope* published by Usborne books, which contains a range of first projects to get both children and adults on the road to become accomplished hobbyist microscopists. At the risk of being repetitive, just in case you did not purchase the book, I'll run you through two projects which are outlined in the recommended work. These are ideal for children and adults but you will need to assist very young children.

A quick bit of general teaching first. All living things, plants and animals, are constructed from cells. Different cells in your body perform different functions. Some animals and plants are unicellular—meaning one cell. That is... they consist of only a single cell. These are usually bacteria, archaea, protozoa, unicellular algae and unicellular fungi. Cells are so small, you need to use a microscope to see almost all of them, although some (very few) unicellular organisms can be seen with the unassisted naked eye.

Cut-away 3D model of a single cell

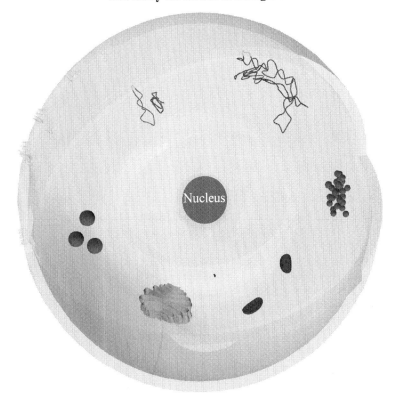

The 3D model of an animal cell shows a variety of special processes inside the cytoplasm and a nucleus which controls and manages the cell. Cells are transparent or semi-transparent and a number of methods are used to stain them to see them more clearly and in greater detail under a microscope.

Fig 1. Unstained cheek cells.

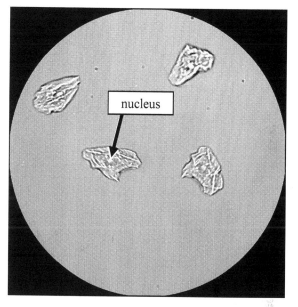

Our first project is to obtain a look at real cells as in Fig.1 above which is, in fact, a still from a video I shot of cheek cells taken from inside a human mouth. (*Slightly blurred here due to grabbing a video frame and over-contrasted by an over-adjusted iris aperture*).

You need a clean glass slide and a clean cotton bud. The inside surface of your cheeks in your mouth constantly shed cells. And luckily for us—they are large and readily seen without staining under a microscope.

Take the cotton bud and wipe it several times gently up and down the inside of your cheek. A lot of cells about to shed or just shed will be deposited onto the cotton wool at the tip of the cotton bud. You now wipe this gently onto the surface of a glass slide and put it under your microscope.

Start with a low power objective and focus the microscope using under stage lighting. You will need to adjust the iris aperture to obtain an optimum view and you may need to scan / search the slide to find a few which have not become too rolled out or distorted. You should be able to identify the nucleus at the centre of the cell, just like in Fig 1. on the previous page.

My student, Pippa, wiping her cheek with a cotton bud. Notice the glass slide at the ready in her other hand.

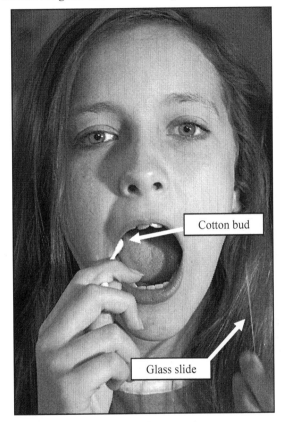

That one was nice and easy. The next one is just very slightly more complicated and involves using a sharp knife, so be careful.

Onion cells
Onions have huge cells, which again are very easy to see. Since we can't stick a whole onion under a microscope, we need to take a thin layer (membrane) from one of the many skins. Part of the skill of a microscopist is learning how to extract and prepare organic material to study under a microscope. Nature doesn't give up her secrets easily. Pictures explain what to do quickly, so...

1). Carefully cut an onion into quarters.

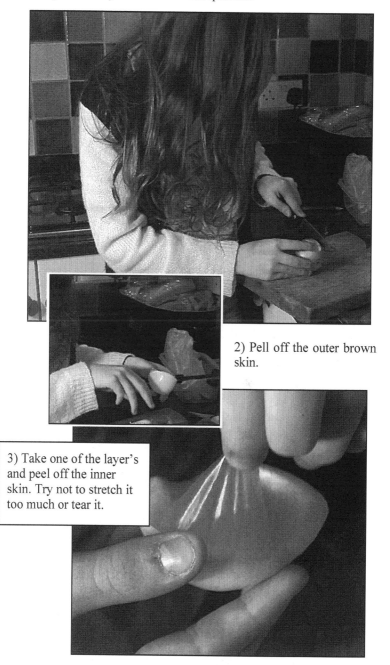

2) Pell off the outer brown skin.

3) Take one of the layer's and peel off the inner skin. Try not to stretch it too much or tear it.

4) Cut a small section from the skin, thus...

(4)

(5)

5) You need to keep it wet so gently lower it into a saucer of water.

6) If you have a small bottle of Iodine (purchased from a chemist or drug store), put a few drops in the water and leave for 10 minutes. Gently run under the tap to wash the iodine from the water. The iodine will have stained the skin and will help you see the nucleus more readily in each cell. If you have no iodine, just carry on as follows. You'll still see the nucleus but it will be fainter. (*Note: no pictures for this step!*)

7) With scissors, cut a piece from this membrane and gently lay it out on the surface of a glass slide. (*Note: no pictures for this step!*)

8) Then use an eye dropper to put a few drops of water onto it.

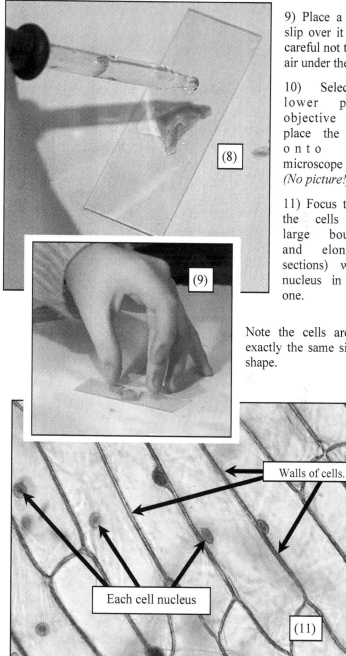

9) Place a cover slip over it being careful not to trap air under the slip.

10) Select a lower power objective and place the slide onto the microscope stage. *(No picture!)*

11) Focus to see the cells (the large bounded and elongated sections) with a nucleus in each one.

Note the cells are not exactly the same size or shape.

(8)

(9)

Walls of cells.

Each cell nucleus

(11)

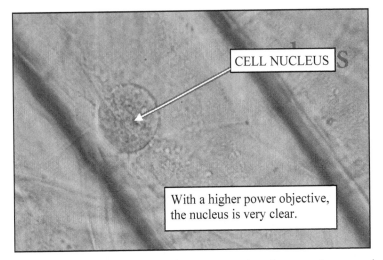

CELL NUCLEUS

With a higher power objective, the nucleus is very clear.

.The last two simple projects are mentioned over and over again in various books on microscopy for beginners for good reason—they demonstrate the easiest way to show you a cell and nucleus without the need for complex fixing, staining, and mounting techniques. There is another way though. Do you recall I mentioned unicellular plants and animals? Many of them live in ponds, streams, and rivers which you'll discover in the next chapter. Plant and animal tissue has to be cut into ultra thin slices and fixed in a chemical mix to preserve their shape and structure, and then stained. Almost all processes in these tiny transparent sections have no (or too little) colour to distinguish them apart. Staining the tissue solves this issue.

To enable a tissue section to be viewed over many years without deterioration, it is mounted onto a glass slide with a cover slip over it, and then sealed. The material used to mount the sections in can be a mix of chemicals—and there are *different* mixes—but all must share something in common. They must allow light rays to pass through the mounting medium in a desired way. This is due to something called the Refractive Index of different substances (R.I.).

R.I. is a measurement of the deflection of light leaving one medium, say air, and entering another medium of a different density, say water, at an angle. A vacuum has the value 1 and the glass cover slip is 1.518. Microscopists have to take account of the mounting medium for different types of specimens because the greater the difference between the respective R.Is of the specimen and the mountant used to affix it to the slide, the greater will be the contrast, and thus the better the image when that slide is examined under the microscope.

Chapter 5. Slide Mounting.

It is becoming increasingly difficult for enthusiast microscopists to source the traditional chemicals used to make preparations to clean, fix, stain and mount microscopic specimens onto a glass slide—something which is done to ensure a specimen can be seen unspoilt for many decades. In the UK, several experienced and long-practicing microscopists share and publish important guides on these processes and the mixtures required. Sadly, many key principles among them have died in the last decade, and although their guides survive, young people in particular find these practices outside of their remit, or simply cannot obtain the chemicals required. Only a minority of more experienced keen microscopists still do their own clearing and fixing, staining, wax embedding (immersing prepared specimens into small wax blocks), and sectioning: cutting thin slices from specimens embedded in paraffin wax.

The General Process

Apart from subjects like whole mounted insects, foraminifera, minerals, and rock sections, most specimen slides would be mounts of plant and animal tissue, insect parts, or micro-organisms. Plant and animal tissue require sectioning, which is the very difficult process of embedding the tissue into molten wax, letting it cool, and then cutting it into very thin sections. The reason for this is that allows a very thin (typically 5 μm *thick*) slice to be cut. The paraffin wax fills the tiny spaces in the tissue structure and helps preserve both its shape and form as well as making it hard and thus easier to cut.

The wax-embedded block can then be sectioned by hand (difficult) or by using a manual or automated microtone (a cutting device). The thin slices undergo subsequent processing before final mounting (fixing) onto a glass slide in a suitable solution to maintain sterility and an appropriate Refractive Index.

Live specimens are often killed in a killing jar using gas which 'relaxes' them. In this instance 'relaxing' means they die without crumpling up in such a way it would make them difficult to study later.

Note: *insects can be embedded whole in plastic resins. This is not normally done for sectioning and study with an optical light microscope, but for educational study or as a novelty: acrylic block key ring adornments for example. Some resin embedding is done for another form of microscopy but this subject is outside the scope of this book.*

A simplified step-by-step guide

Fixing

Small tissue blocks or whole subjects (plants, insects, organs, etc.) are chemically fixed. The chemical used binds and cross-links some proteins, and denatures other proteins through dehydration. This hardens the tissue, and inactivates enzymes that might otherwise degrade the tissue. It also kills bacteria. Traditionally, the fixative most commonly used is a 4% aqueous solution of formaldehyde, at neutral pH.

Dehydration and clearing

Paraffin wax is not soluble in water or alcohol. All cells, tissue, plants, animals etc, are full of water which needs ultimately to be replaced by wax to make a wax-embedded block of the specimen. To achieve this, the water must be removed and replaced with xylene—a paraffin solvent. This cannot be done in one go. First the water has to be replaced by alcohol, a liquid which is miscible (capable of being mixed) with xylene. The specimen is subjected to a series of increasing concentrates of ethyl alcohol baths, where each subsequent one contains an increasing amount of alcohol from 0% to 100% concentrations). Once the specimen is 100% saturated with ehythl alcohol, it can be bathed in xylene to drive out the eythyl alcohol.

Wax embedding

The tissue is placed in warm paraffin wax, and the melted wax fills the spaces that used to have water in them—now xylene. After cooling, the tissue hardens, and can then be cut into thin slices (sectioned).

Sectioning

Very thin slices are cut from the wax block using a microtome.

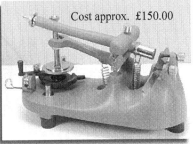

Cost approx. £35.00

Cost approx. £150.00

Examples of hand microtomes.
Above is a rocking type.

Staining
Staining in one or more dyes is normally required to help differentiate different processes in the specimen. Alas. most staining solutions are aqueous but our wax embedded sections contain no water. The wax has to be dissolved and replaced with water—rehydration. Effectively, this is the stage above *Dehydration and clearing*, but in reverse! The sections are passed through xylene, and then decreasing strengths of alcohol (100% to 0%), and finally water. Once stained, the section goes through *Dehydration and clearing* all over again until the section is completed saturated in xylene.

Mounting
The specimen is then mounted on a glass slide in a mounting medium such as Canada Balsam, although there are many others. The mounting medium holds the specimen in place between the cover slip and the slide. A coverslip is placed on top, to protect the sample. Evaporation of xylene around the edges of the coverslip dries the mounting medium and bonds the cover slip firmly to the slide.

I think you can see by this that permanent slide mounting can be a sophisticated and painstaking endeavour even for the most seasoned amateur.

Note: *Xylene is now considered toxic and is not readily available easily for the amateur microscopist.*

A DVD to show the technique of wax embedding
At the risk of seeming like I am constantly plugging resources from Brunel Microscopes Ltd—which is no bad thing, as they are a commercial friend not only to me but the entire amateur microscopy community, quite a rare thing now in a multi-corporate often ruthless corporate environment—they sell a DVD explaining how to do wax embedding. Well worth the cost if you want to get involved with it.

Paraffin Wax Processing DVD: An excellent DVD presentation of the entire process of specimen selection and paraffin wax processing to the final section cutting with either a rocking or rotary microtome. This is the information that is needed to produce stained sections from almost all tissue types. Invaluable information. 40 minutes duration with detailed sound commentary. Price £12.50

Purchase via the web here:
http://www.brunelmicroscopes.co.uk/microtomy.html

Alternatives

Clear Nail polish

Nail polish can be used to seal the sides of the coverslip when using aqueous mounting media. It can also be used directly as a mounting medium. The specimens must first be dehydrated in alcohol and can then be directly mounted (without xylene) in nail polish.

Glycerol jelly

This can be a most difficult mounting medium to use, but sometimes there is no other satisfactory alternative to an aqueous mounting medium. Water-based mounting media are useful for making permanent mounts of water organisms, algae, protozoa, etc. Glycerol jelly is commonly used to preserve pollen samples.

Other 'easier to obtain' Mounting Mediums

Distilled water, glycerin, sugars (karo and fructose), gum arabic, gelatin and PVA make very economic alternatives to Canada Balsam Together, the selected products provide a selection of aqueous media, two of them liquids (antiseptic water and glycerol), with the others—solids, and also one easy to use synthetic resinous medium, NPM (Nail Polish Mountant).

Various mixtures and techniques are employed. I would refer you to this work by Walter Dioni for a comprehensive learning of safe mounting media and techniques. It is available as a paperback and on kindle from Amazon:

SAFE MICROSCOPIC TECHNIQUES FOR AMATEURS Slide Mounting [Paperback]

Paperback: 102 pages
Publisher: CreateSpace Independent Publishing Platform; 1st edition (31 May 2014)
Language: English
ISBN-10: 1499746512
ISBN-13: 978-1499746518
Product Dimensions: 20.3 x 13.3 x 0.6 cm

Mounting in fructose

A good method for mounting some specimens without getting involved with more volatile chemicals, and thus an ideal method for amateurs, is using a type of sugar as a mounting medium. The following step-by-step guide can be tried by most people quite easily.

Using FRUCTOSE SUGAR as a mountant

This is our magic ingredient which helps to make slide-making more accessible to younger people or, in fact, even to long-standing Microscopists. We are going to make up a stock solution ready to use for making loads of slides. Remember though that because we are going to make slides using a SAFE method, we are going to skip a couple of important processes which would require the use of chemicals not quite as safe to use for youngsters. This will mean that some specimens will not be perfectly mounted for absolute clear 'viewing'- but we should end up with slides acceptable to the beginner, who may then wish to move on and adopt the processes omitted here to refine his or her future slides!

Materials - basic needs!
You need some glass slides. After all is said and done, you can't escape this requirement. Glass has a 'see-through' trait despite the fact that it is considered a material that can easily cut and cause injury. You just have to be careful how you handle it. You also need cover slips. These wafer thin 'slips' of glass will be used to cover the specimens you place on the larger glass slides. I'm currently using cover slips which are 22mm square, and glass specimen slides 76mmm long, 21mm wide, and 1mm thick (3 x 1 inch).

Where do you get them from?
Well this is the part where you have to contact a small company that sells them. You can't pick them up from your local shop so you need to order some. Many places on the internet will provide them.

A few more materials
Here's a little picture of some extra things you need to make your first

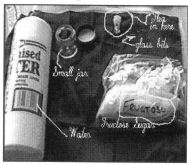

slide. There's one more thing, not shown here, that I'll explain to you in a moment... but first of all, let me talk you through these items. Most of them are easy to obtain from around the house or a local store so don't panic about how you're going to get them. It's easy!

Water
This is not ordinary water! It is De-ionised water - the type used in car

batteries and electric steam irons. You can normally buy it cheaply at a garage or a motor shop or from a super-store. It is perfectly safe in every respect and differs only from normal water by the fact that it is very pure (free of minerals) and very clear. We will use it to help make a really safe mounting solution. I hope you recall this is used to support and hold the specimen in an air-free condition onto the glass slide.

Glass bits
Not so safe! I got my glass bits by breaking up a cover slip into smaller pieces. These will be used to hold the cover slip in position just above the specimen, effectively acting as 'spacers' between the glass slide and the cover slip. I'll show you as we progress how you can use something safer than bits of glass to do this.

A small glass jar
The jar is very small. We'll use it to make up and hold the mounting solution we make. When we make the solution, we will create enough to make loads of slides and since we won't use it all at once, we need a place to keep it.

   You could wash out one of those small jars used to hold fish or meat paste. If you don't have one, what about one of those tiny jars you can buy jam in - you know - they look like samples rather than jam jars. Take a look in the larder or fridge and see if you can find a small container you can adapt. You must wash it out properly and make sure it has an air-tight cap or lid.

A specimen
What are you going to mount? I managed to find a flea on my dog and popped it in a jar while I thought about how I would kill him without hurting him. This is exactly how I found him (her?). By luck, he just died naturally. I opened the little container the day after I put him in it, and he was just laying there dead.

Some of you may not wish to kill little animals, so why not choose something like some pollen from a flower, hairs from a nettle, or anything else. The smaller the specimen, the easier it will be to make your slide. I'll show you how I mounted pollen and some plant hairs too as part of this lesson so you can see how easy it is.

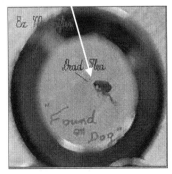

Fructose sugar

Our first shared secret: Fructose! This is a type of sugar found naturally in foods like fruit. Natural Health food shops sell it by the box quite cheaply as an alternative to cane sugar. It is useful to us microscopists because it provides a clear sticky liquid, when mixed with the right quantity of water, which by complete coincidence is ideally suited as a mounting medium for a lot of microscopic specimens. It's safe, non toxic, and cheap! It will not enable you to make a slide that will last for ever, but it will allow you to make one which will last at least a few years and maybe much longer with little deterioration.

Still more bits...

We need a few more items. The first is a bottle of nail varnish, the type used by women to paint their nails. You can borrow some from your mum, girlfriend, wife etc., if you don't normally use it yourself. I went out and bought some CLEAR nail varnish rather than be stuck with a coloured one. You will use this to seal the edge of the cover slip to the glass slide.

If you do not wish to use a broken cover slip as spacers, there is a better way if your specimen is nice and small and flat. You need to purchase some PVA glue from an ART or DIY shop. This is that milky -white type glue which dries clear. It's normally used in collage work to stick bits of card and paper together and is quite cheap to buy.

You will need a small artists type paint brush too. One with a fine narrow point. The one I use has written on it: 000 DALER HP 70, if that helps you at all. You will need two of these if you wish to make a better job of making the slide by using enamel paint: see below!

Non-essential

One more thing which you don't absolutely need (but it helps to make a nice job) is a small tin of ENAMEL PAINT. I'm using some stuff left over from building Airfix plastic models. It's made by Humbrol and comes in a very tiny can and in many different colours. I recommend black or very dark blue. This is used in the very final stage to paint over the narrow band of nail varnish, thereby providing a stronger seal and great protection where the cover slip edges meet and seal with the glass slide.

Working space

Once you have all the items ready, you're ready to start. It's best to clear an area of a table so you have a clean space to work on.

Pour some Fructose sugar into a clean dry glass jar. See overleaf: (1) & (2). Ideally, you want to fill the jar two thirds up to the top. Pack the sugar down so it is firmly taking up the space in the lower part of the jar. Take a pencil or felt-tipped pen and mark the level of the sugar in the jar. You do this by drawing a small dash on the outside of the jar exactly where the top of the sugar reaches to.

Gently pour some de-ionised water into the jar. The sugar is

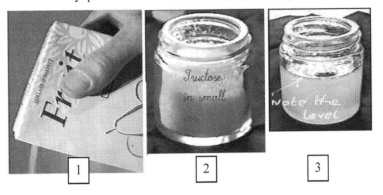

very soluble and will start to dissolve immediately causing the level to drop. You must keep adding water to maintain the level to your original mark (3) When most of the sugar seems to have dissolved, you will notice the liquid is not yet clear because it retains crystals of Fructose which still have not dissolved. Put the cap back on the jar and leave it in a safe warm place overnight. This will allow time for all the sugar to dissolve and you should end up with a nice clear liquid.

(4) This is how it should look when you fetch it out in the morning—clear.

Important extra step!

You can use the Fructose in this strength (as mixed above) but I have found it easier to mount the specimens I've done so far in a slightly weaker solution. Some things tend to 'float' on top of the thicker solution and if you end up getting air bubbles in the drop you use on the slide, they are more difficult to get out.

I recommend you add more water to the solution at this stage to raise the level to half way between your mark and the top of the jar! Put the cap back on firmly and turn the jar upside down and then right way up a few times to gently merge the water with the syrupy solution.

Leave for a few hours before using it to ensure the sugar stock 'thins' out throughout the jar. When you have a ready supply of Fructose Mountant, and all the other materials ready to hand, we'll move on to make the slide!

The theory

To make a slide, we need to create an extremely thin well or cavity on a glass slide in which to place our specimen and our Fructose mounting solution. There are many ways of doing this. We could, for example, use pieces of a broken cover slip and stick these onto the glass surface of a slide. Ultimately, the cover slip would sit-down onto these pieces - which prevent the specimen from being crushed.

I'd like to show you a better and safer way. First of all, take a look at (5) which shows a sketch I made for you. Does this give you an idea of how we will build a well on the slide?

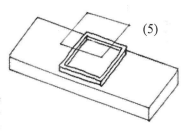

(5)

We will use PVA glue to create a thin wall that will run just inside the edges of the cover slip. PVA glue is ideal for this. It can be painted onto glass and dries very quickly. Some specimens, like my flea - for example, will require a higher wall to ensure the cover glass does *not* sit on the specimen itself. We can build our well to be as deep as we like simply by adding layers of PVA glue after the previous one has had time to dry.

Let's do our PVA glue wall now together. Lay your slide on top of a cover slip, ensuring that the slip itself is centrally placed under the glass slide. By doing this, we can see where the edges of the cover slip will be when - in a later stage - we fix the slip on top of the slide.

Take a fine paintbrush and moisten it. Maybe you should put some water in an eggcup and keep it close to hand for this purpose. Now open up your PVA glue, or lightly squeeze some out and gently dip the tip of the brush into the milky white liquid. You only need a tiny amount of glue. Do not overload the brush. With a steady hand, paint a fine line onto the top of the glass slide just inside where you can see the edges of the cover slip beneath it.

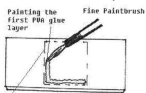

Painting the first PVA glue layer   Fine Paintbrush

Cover slip placed under slide

It's important to try and maintain a continuous line of equal thickness and depth all the way round. You can remoisten the brush and reload it with glue as you go, but remember to shake off excess water from the brush before loading it with glue. If you have problems holding the glass slide still and in position on top of the slippery surface of the cover slip, try the following instead.

Put a glass slide on a sheet of paper and draw a line around it with a pen or pencil. Place a cover slip centrally in the rectangle drawn and trace around this with your pen. Now you can use your drawing as a template by laying the glass slide on top of it instead of the cover slip when painting on the PVA glue.

After you have completed the PVA line, put the slide to one side for the glue to dry. It will be ready quicker if you put it somewhere warm. I sit mine on top or near a central heating radiator and they dry in a few minutes. You'll know when the glue is dry because it changes colour from white to nearly transparent.

You may wish to make several slides in your first batch. If so, prepare a few more and put these aside to dry.

Take a look at the specimen you are going to use for your slide. Is it as big as a flea? In my first batch, I made slides of a flea, pollen grains, hairs from a leaf, and a mould found on bramble leaves. The first two required me to build up a PVA wall in several layers to ensure the specimen would 'sit' down properly inside the well. If you fail to put on enough layers, you won't be able to get the cover slip to settle down on the dried glue properly!

Preparing the specimen
Variations
There are several ways you can prepare subjects for immersion in the Fructose. One way is to put a tiny amount of the solution into a clean small dish or egg cup. Put your specimen into this, being careful not to introduce air bubbles, and leave it for a few days for some of the water to evaporate - leaving the solution much thicker! If you wish to try this method, stand the eggcup or small dish between two pencils and rest an up-turned tumbler or glass over the eggcup but resting on the pencils. This will allow air to flow through but prevent dust from falling into the solution! It is a good idea to at least soak your specimen in Fructose for a night so that the solution can seep into its structure.

When you have been an amateur microscopist for a while, you will learn different ways of using chemicals to help clean and fix the specimen prior to mounting it in the fructose solution. I am deliberately skipping this important step here because it involves using chemicals which are normally highly inflammable or slightly toxic. As this is your first slide-making try, I want to keep the process as simple as possible.

Another method of preparing the specimen for the fructose is to place it into some clean water. For example, I plunged some leaves from a bush into boiling water and left them in there overnight. This helps to force air out of the plant which, if not done, would prevent fructose mountant from seeping into the leaf's structure. I scraped a leaf's surface with a razor blade to remove the hairs into a saucer containing a few drops of fructose solution diluted heavily with water. Lets use this as my example for the best way of proceeding.

This is what you do:

Using a smooth implement like a knitting needle (I use a slim glass rod), insert the tool into your jar of Fructose solution. Remove it and bring its point down into contact with your glass slide, ensuring the point of contact is centrally inside the 'well'. Let the Fructose drop slip down onto the slide.

You need to judge how big a droplet to leave on the slide. You should be aiming to lay down a drop big enough that when you finally place a cover slip on top, the solution will spread out and fill up the PVA-glue formed well. In the long run, trial and error will enable you to judge this. You simply repeat the action to build up a bigger drop of solution in the well.

Work carefully and precisely trying to avoid introducing air bubbles into the droplet. When you think you have enough solution in the 'well', position your specimen on the slide by pushing it gently into the fructose droplet. I used my glass rod to pick up a drop of watery solution (containing plant hairs) from the saucer and introduced this to the Fructose droplet on the slide.

Lowering the lid

Take the cover slip and, either using tweezers, or holding it gently between index finger and thumb, breathe two or three times rapidly onto its surface. This is to moisten it which will help the Fructose to run without trapping air when we lower the slip onto the slide in a moment!

Working quickly, lower the cover slip down onto the droplet, keeping the edges aligned with the PVA-glue on the slide. Just before the Fructose droplet comes into contact with the underside of the cover slip, release the slip - letting it float down onto the PVA glue walls. The Fructose should spread out across the underside of the slip and fill the 'well' without leaving any air gaps.

Check! If the slip is not resting squarely on the dried PVA glue walls, gently slide the cover slip sideways until it's properly positioned.

If you find you have used too much Fructose mountant - causing it to leak out onto the glass slide itself - use moistened cotton buds or tissue paper to gently clean and wipe the spilled fructose away. Ensure you leave the area around the cover glass as clean as possible.

Right, how did you do?

You may find you have air bubbles trapped under the cover glass. If these are very few and not all over the specimen itself, leave it be. If you have a lot of them, it might be best to recover the slide, cover glass, and specimen by placing the whole thing in a saucer of warm water. You can clean the glue from the slide by peeling it off. Dry and clean the glass slide and slip and reuse it again for another go.

This is what may be causing air bubbles to form at this stage:

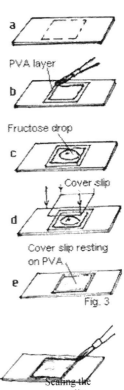

PVA layer

Fructose drop

Cover slip

Cover slip resting on PVA

Fig. 3

Too small a drop of Fructose used.
Cover slip was not moist enough.
Cover slip was not clean enough.
Fumbling around as you let the cover slip go.

Before I move on to the next stage, here is a recap of what you needed to have done to get to this point:-

(a) Position the cover slip under the slide to act as a 'see-through' guide.
(b) Brush on layer/layers of PVA glue.
(c) Place Fructose Drop and specimen in the well.
(d) Lower the cover slip (after breathing on it).
(e) Ensure the slip rests on the 'well' walls squarely.

If you are satisfied with your slide or slides up to this point, put them somewhere warm for a few hours or better still, leave them somewhere dust-free (remember the pencils and tumbler tip) for a day or two. However, if you are in a rush - like me - just make sure the area around the cover slip is dry, free of water and spilled fructose, and carry on as follows.

Sealing the cover slip

Sealing the slide with nail varnish

You now need to use the nail varnish to seal the cover slip to the slide. Nail Varnish comes with its own little brush. Use this by wiping off the excess when taking the brush from the bottle. Dab a drop of the varnish on the very edges of the cover slip in each of its 4 corners. Each drop should be half on the cover slip and half on the glass surface of the slide itself. The idea here is to fix the cover slip to the slide so that it doesn't move around in a moment when we paint right around the edge of the cover slip with the varnish.

Take care when applying the dabs so as not to move the cover slip or spill varnish onto the central area of the cover slip. Put the slide to one side for 5 minutes to let the varnish dry. Be sure to return the brush to the bottle right away so the varnish doesn't dry on the brush itself.

When the 4 spots have dried on the slide, carefully run the nail varnish brush all around the edge of the cover slip. The edge of the cover slip and the surface of the glass slide up to about 1/8 inch around the slip should be covered in one continuous line. The idea here is to completely seal the edge of the cover slip where it meets the glass slide.

Put the slide to one side and let the varnish dry. It will take about 5 minutes in a warm room, and then you can apply another layer on top of the first one. Work quickly so the solvent doesn't melt the first layer as you sweep the brush over it!

Here's one I made so you can see just how wide a strip of nail varnish to apply. Look at it carefully. Can you see all the elements we have spoken about so far? You should be able to see the PVA glue beneath the cover slip. To do a proper job, write out a tiny label and sellotape it to the slide as I have done here. ⟶

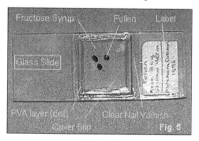

Enamel Paint

At this point, your first slide is finished but if you want to do it the right way, you should apply a layer of enamel paint once the varnish has dried. The paint must be painted on as you did the varnish.

Here's one I did (next page).

Notice that I didn't apply enough Fructose Mountant in this one which has left some air bubbles under the cover slip. They are well out of the way of the specimens in the middle, so I decided not to recover the slide but keep it instead.

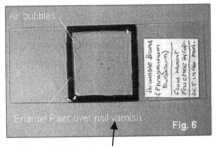

A few air bubbles in my one

After all that, I guess you want to see what my slides look like under the Microscope. Okay then... pick from these 2 that I made. (*See below*). Alas, I can't show you the flea one because I broke this accidently one night while I was looking at it under the scope: *my first broken slide!*

This is a low power shot, and then a higher powered one taken att he microscope of hairs found on a New Zealand Daisy Bush, Olearia haastii.

It is seen here in polarized light. I scraped the surface of a leaf and deposited the 'scrapings' into a drop of Fructose Solution.

Apologies for small image size,

but it is the process which is important for you, not my result.
This is my favourite one (*below*). It is commonly known as Bramble Brand, Phragmidium bulbosum. It is a mould found on Bramble leaves and these are the spores scraped from one such leaf. You can often see tiny brown specks covering bramble leaves, each speck is 'raised' from

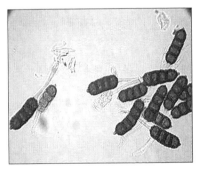

the leaf surface. Now you know what it looks like under a microscope. Good News. I just decided to check a few slides I fructose mounted back in 1997, over sixteen years ago. I'm amazed! No deterioration whatsoever. It means fructose can be used for long term mounting. Here are some completed along with a box you can make out of old

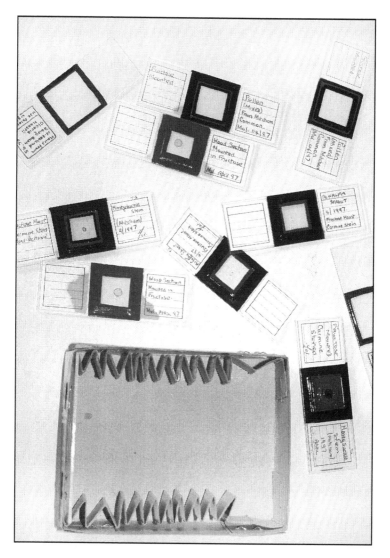

cereal boxes to keep them in. See how I used two strips of cereal box and folded them so the slides can slot into them (above).

Dry mounting

Dry mounting is one of the oldest methods of preparation to be used in microscopy. It is particularly suitable for Foraminifera as well as for chemical crystals, clothing fabrics and dried out objects such as very tiny insects. There are several methods but I'll explain the simplest and cheapest one. In general terms, dry mounts are used to contain the specimen in a small well, sealed with a cover slip on the glass slide. First buy some double sided sticky poster foam mounting squares.

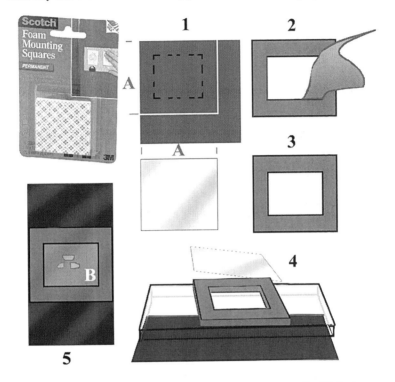

Cut from one of the squares a square equal to the cover slip [1] & A.
Cut a square inside the foam and pull it out [2].
Attach the sticky surround foam to the glass slide [4].
Put your specimens in the well (glued or loose) [4].
Push cover slip down onto sticky foam square [4].
Cut and glue black paper to back of slide or paint black with enamel paint. Note: in [5] I've shown the area of hole under the cover slip as a lighter colour whereas it would look black really. There you have it, cheap, simple, and easy to see specimens with over stage lighting. This method was sent in to our Micscape Magazine by *Jim Benko, USA*. Just note there are other ways in which under-stage lighting can be passed up under the slide through a transparent ring around the dark background such that it reflects back down to illuminate the specimens. To see this method, go online to our web site at:

http://www.microscopy-uk.org.uk/mag/artoct01/bdmount.html

Chapter 6. Pond Life—A Quick Dip

An introduction.

If you could climb into a microscopic submarine and travel around or across a small well established garden pond, you will have a journey which surpasses Jason and the Argonauts and the star ship Enterprise, and you would witness living creatures and plants more incredulous than dinosaurs, ET aliens, and every other macro-sized life form on the planet.

But you can't!

So the next best thing is to bring them to you and study them under a microscope. Not only is the study of pond life (fresh water) one of the most engaging branches of microscopy, it is also critically useful in controlling and managing an important environmental issue: clean, drinkable water! This is the one resource on our planet, maybe throughout the universe, which enables life to be in the first place. Without clean water—there is no sustaining of life. The interesting point though is we do not study the water, but the tiny things swimming in it. And what flourishes in this miraculously simple substance can define whether it is a lake of toxic poison, or a water garden of Eden. There is a great myth regarding the Amazon rain forest concerning the way it is purported to be the lungs of the earth. This is utter rubbish! Where they do serve life on the planet well is in the fact that the forests, and the Amazon is a main contributor to this, provide a filtering system for the atmosphere—they are natural air scrubbers. They do not provide any real additional oxygen because they absorb as much as they give out. The two single most critical elements on earth, without which all life would be extinguished, is water and the microscope plants living in it: *algae*.

Algae have photosynthetic machinery ultimately derived from cyanobacteria that produce oxygen as a by-product of photosynthesis. Effectively, they absorb sunlight, produce oxygen, and reproduce. They exist in all bodies of water including our vast oceans. They also form the basis of the major food chain since they are 'loosely' the first lower order main physical living form providing nutrients to larger microscopic animals—rotifers, for example—and to non microscopic life forms such as fish. Without algae and other photosynthesizing microscopic forms, there would be insufficient oxygen in our atmosphere to support life in the abundance we witness today.

No small wonder then that so many variant forms of microscopic life thrive in our oceans, rivers, and ponds.

A single well balanced pond can be regarded by the ever exploring amateur microscopist, as a complete world or bio-system which can provide a lifetime of fascinating study and learning.

Studying water samples

Samples of water can be taken from different areas and depths of the pond. Drops of the water are deposited on glass slides using an eye dropper or pipette, and a cover slip is gently put down onto the water. If you wish to study the slide for longer periods, a tiny sliver of Vaseline is put around the edge of the cover slip to provide a temporary water seal with the glass slide, and prevent it from drying out too quickly.

Very fast moving specimens can be slowed down by adding gelatine or a similar transparent thickening substance to the water drop to increase its viscosity (note: best to add gelatine to the jar of pond water sometimes). Gelatin can be purchased from the cooking/baking section of most supermarkets either as a liquid or as a sheet you can melt in hot water.

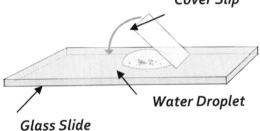

Cover Slip

Water Droplet

Glass Slide

Above. This is the correct way to place the cover slip over the water droplet to avoid air bubbles. Position the cover slip as shown and then let it fall in the direction indicated. Some surplus water will spill out from under the cover slip. Creep the edge of a piece of paper towel or tissue up to the cover slip edge to absorb and remove the spilt water.

Remember you will be studying living micro-creatures and plants so there is no staining involved. Various, more sophisticated

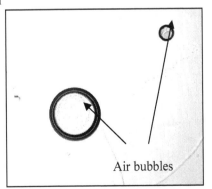

Air bubbles

lighting techniques, which invariably involve additional microscope attachments can be employed to gain greater insight through improved illumination and separation of details in the subject observed, but I suggest you leave this aspect until you feel you might like to develop your skills further.

If you fail to put the coverslip on in the way described, you are likely to introduce air bubbles into the sample which will spoil viewing of the subject.

Breeding and culturing

It's possible to extract a species of micro-life, say a specific rotifer, and introduce them into a jar of water containing a food source which promotes growth and reproduction. A few microscopists do this which then allows them to sample, view, or photograph them at leisure. Batches can be cultivated and sent to people wanting to introduce them to their ponds (great for helping get rid of green water in new ponds), or smaller samples can be sent to other people wishing to study live specimens.

This can be done very simply by sterilising a small glass jar through heating it in an oven, then after it has cooled—by half filling it with de-ionised water, and by putting a few grains of rice into the bottom of the jar. Introduce a few drops of your pond water sample ensuring you have some rotifer (or other critters of choice) in the drops. The rice breaks down slowly through bacterial action providing a steady supply of bacteria for the rotifer or other life form to feed on.

Non-pond water

Nature has developed remarkably well-adapted life forms. There are many microscopic forms which can live in water and should that water evaporate or disappear—they can transform to a desiccated state (*dried out*) and continue to exist in that state for considerable lengths of time until the water returns, and they revert to their active moving, eating, and reproduction activities. One of my favourites can be found in most bird baths, hand-holds in man-hole covers, or other places which form small pools of water through rainfall and then dry out until the next downpour. The one below is red (rust-like) in colour and you can detect it in dried out bird baths as red dust-like deposits. Scrape some off, take the sample home, put a few specks on a slide and add a few drops of water. Under the microscope, watch as it miraculously seems to come alive again.

Rotifers

Although microscopic, rotifers are animals. Some give birth to live offspring. With over 2000 species of them, some living within their own shells, others living free—they are one of the most visually interesting creatures to look at. Rotifers were some of the first microscopic life forms ever seen with a microscope back in the 17th century and were named wheel-animalcules, due to the fact that many seemed to have spinning windmill like appendages atop their heads (one each side of

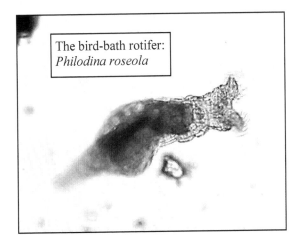

The bird-bath rotifer:
Philodina roseola

their mouth), which propelled them and drew food into their mouths. A pair of rock hard jaws can be seen inside the animal pounding down any stray algae drawn into them through the whirling cilia (fine hairs). Some build nests or create protective shells to live in from minerals found in the water. The rotifer below shows that the main body is

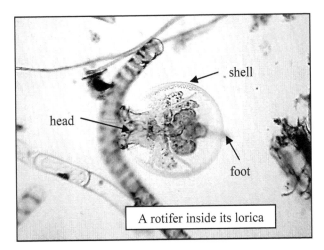

shell

head

foot

A rotifer inside its lorica

protected by a lorica (shell) allowing the head to come out, and with a hole beneath for a single foot to extend through and cling onto plants to stabilize itself when feeding in a stationary position.

Paramecium
Commonly known as the 'slipper animal' as it's shape often resembles a slipper. Paramecium is one of the micro-organisms used as a typical example of microscopic life in school biology. Most students will

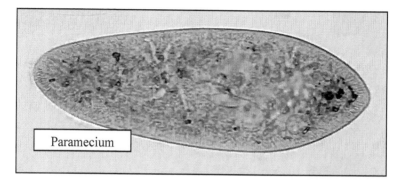
Paramecium

encounter this wonderful lively creature.

Diatoms
Diatoms are a major group of algae, and are among the most common types of phytoplankton. Most diatoms are unicellular, although they can exist as colonies in the shape of filaments or ribbons. A unique feature of diatom cells is that they are enclosed within a cell wall made of silica called a frustule. These frustules show a wide diversity in

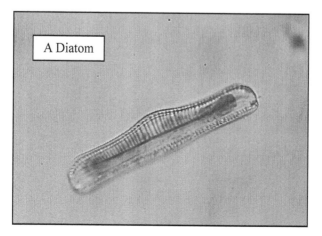
A Diatom

form, but are usually almost bilaterally symmetrical, hence the group name. The symmetry is not perfect since one of the valves is slightly larger than the other allowing one valve to fit inside the edge of the other. Diatom communities are a popular tool for monitoring environmental conditions, past and present, and they are commonly used in studies of water quality. Less well known is that are often looked for in forensic analysis where a dead corpse is discovered and suspicion exists of death by drowning. The body can be searched to see if diatoms were ingested or taken into the lung. Their silica shells do not decay readily and remain long after the death of the individual. Discovering diatoms in lungs or bone marrow can indicate if the 'victim' was submersed as a living person or had died beforehand. Also, the diatoms can be matched to samples living in the water where the person is suspected to have drowned.

Micro-life to excite the very young.
A low power stereo microscope suited to very young children is unlikely to excite their imaginations through observing truly microscopic animals and plants. Twenty times magnification is not quite enough. However, there are other bigger things to look at. Live Daphnia (freshwater shrimp) can be purchased from many garden centres specialising in garden ponds for a few pounds. Water fleas, water skimming beetles and flies, tiny worms (nematodes), water snails, mosquito larva, water beetles and many other creatures are large enough to provide magnificent and spectacular displays under a very low powered microscope... and they are much easier to obtain as you can see them.

Summary
I could fill the rest of this chapter, the remaining pages of this book, and create 10 more just showing pond life creatures and plants, but I won't. What I have done is to introduce you into a vast area of discovery and the main one which amateur microscopists enjoy. So many other books and internet resources exist, which are easy to source, focusing on fresh water and salt water (oceans) microscopic organisms. The micro-animals and plants have been written about and drawn for over 150 years, and photographed and filmed for decades.

What has never happened though is that no imaginative established or budding block-buster feature film maker, including Disney, has ever capitalized on their existence. But more on that later. Let me show you just a few more of the beautiful and exotic micro life forms you can witness for yourself by studying pond life with a microscope.

Volvox

These are wonderfully graceful forms to behold spinning slowly like semi-transparent orbs through the water. Volvox are spherical colonies of green cells clinging to a semi-transparent hollow ball of mucilage. A

Volvox image— a frame from a video by Ken Jones.

single colony may consist of over 500 cells, each one with a tiny pair of whip-like tails (flagella) - and all cells undulating their flagella in unison, propelling the colony through the water. Very large colonies can exceed 1 mm in diameter and are easily visible to the naked eye. Many will be found to contain daughter cells, and sometime even grand -daughter cells in various stages of development within the hollow interior of the globe.

Bryozoans (See image opposite page—top)

These beautiful bryozoans look, at first glance under a microscope, like delicate fairies of the deep. This genus even has a name - Plumatella repens - which conjures up the 'sugar PLUM fairy' imagery of folklore and children's books. With translucent white tentacles, waving like wings in a breeze, they are a wonderful sight to greet the amateur microscopist. Bryozoans are common animals, but due to their plant-like growth, they are often overlooked by the casual observer. Even with moderate magnification like an 8x lens, their structure and fine mechanisms become apparent. These animals are known as zooids within the colonial mass, which is termed as the zooecium. Each animal is composed of two key parts; a body section containing a gut and a highly specialized U-shaped structure surrounding the mouth, called the lophophore - it is this which bears the wispy tentacles. The

71

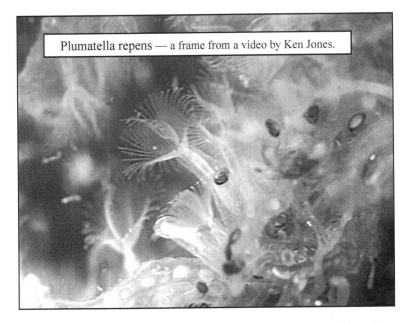

Plumatella repens — a frame from a video by Ken Jones.

tentacles are ciliated - that is, they have many fine hairs along their surface, which undulate in unison, creating currents in the water to draw algae and protozoans (small animals) into their mouth. The tentacles can be withdrawn and retracted when the animal is not feeding. These images are single frames taken from a low resolution (by today's standards) video in 1994 filmed by Ken Jones in the UK who kindly gave permission for me to use his fine work.

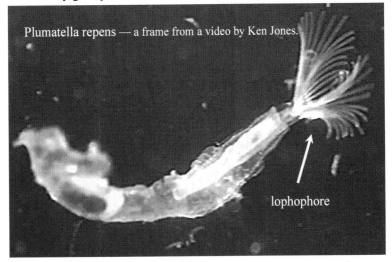

Plumatella repens — a frame from a video by Ken Jones.

lophophore

Chapter 7. A Very, Very Brief History

Before closing this section of what I consider as traditional amateur (or enthusiast) microscopy, it might prove of interest to quickly run through the history of this branch of scientific discovery and evolution. If you consider rocket science, in a modern sense, started circa 1943— just 70 years ago, and then understand that the first real microscopes were used in a rudimentary form from the late 1500s, then you can see why another area of enthusiast microscopy also exists, namely that one of its history over the last 400 or so years.

The Romans and probably civilisations before them were aware that water droplets deposited on leaves and stones seemed to magnify the surfaces beneath. The Romans had glass makers who made bean-sized glass artefacts which were thick in the middle and thin at their edge. They realised that when these 'beans' were held over objects, they exhibited a magnifying effect. They called these glass artefacts 'burning glasses' or magnifiers. The word lens is derived from the word 'lentil' which these early magnifiers eventually became to be called as they resembled lentil beans in shape.

Spectacle makers were producing lenses to be worn as glasses after the end of the 13th century and certainly simple magnifiers of a single glass lens were being used more commonly in the later 1550s, often to look at small insects and fleas. These early magnifiers were

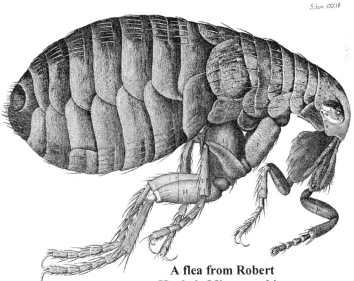

Schem XXXIV

A flea from Robert Hooke's Micrographia.

called 'flea glasses'.

During the 1590's, two Dutch spectacle makers, Zaccharias Janssen and his father Hans started experimenting, putting several lenses into a tube. They realised an object near the end of the tube appeared to be greatly enlarged with a magnification greater than a single glass magnifier.

A draper in the Netherlands (Holland), Antonie van Leeuwenhoek, is credited with creating and employing the best first microscope of that time (1600s). He discovered red blood cells and

Hooke's compound microscope

spermatozoa in the mid 1600s Although his microscope was not the easiest to use, on 9 October 1676 Van Leeuwenhoek reported the discovery of micro-organisms.

An Englishman and curator at the Royal Society—the first truly scientific organisation based in London—Robert Hooke, published a book of his own studies called Micrographia which had a huge impact, largely because of its impressive illustrations. (See page 73). If you would like to know more about Robert Hooke, the single person I regard as the father of modern science, you can go on the internet and look up my article about him here: **www.microscopy-uk.org.uk/mag/ artmar00/hooke1.html**

Robert Hooke, apart from his many other discoveries and inventions, devised one of the earliest compound microscopes of his time. The optical light microscope used both by amateurs and professionals today is still based in part on this early design, but with improvements and some extras.

The Victorians
The heyday of enthusiast microscopy was undoubtedly the Victorian period where many slide makers hand-made beautifully presented slides for wealthy Victorian families to muse over. The Victorians were

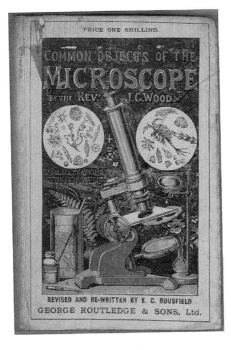

a curious and science-pioneering people and a lot of people interested in microscopy history today tend to focus on this period.

Illustrated books and hefty volumes like Gosse's Evenings at the Microscope sold well, but not everyone wanted such large works, and there were plenty of smaller books for less exacting readers. The Rev. J.G. Wood's Common Objects of the Microscope sold prodigiously and was still going strong (posthumously) 36 years after it first appeared, retailing in the early 1900s for around the same price as a single microscope slide. Circa 1850s.
(Extract and image by Peter B. Paisley. Sydney, Australia. Extracted from Micscape Magazine:
www.microscopy-uk.org.uk/mag/artoct10/pp-aquatic.html)

Here is a wonderful site showing Victorian slides:
http://www.victorianmicroscopeslides.com/

And another on Victorian & Edwardian Microscope Makers
http://microscopist.net/

Photomicrography—taking photos of subjects under the microscope—was underway by 1839 and continued to advance right through into the 1900s. This certainly would have added greater interest in the world of the very small and the interest of the public. But I propose it is likely that the invention of radio and the arrival of radio broadcasts circa 1920 started the slow decline of wide interest in amateur microscopy which worsened with the advance of electronics, television broadcasting—and continues today due to the wide diversity of other novel distractions.

Yet, this should not be happening now. This sweep through advancing technologies in the last 100 years have taken the current generation to a starting point of a whole new era of amateur microscopy and wonderful fresh tools at their disposal to advance microscopy in exciting ways. We have arrived in the 21st century with a long and profound history behind us, but we have also arrived at the entrance of an hitherto unimagined star-gate to a whole new world where we can exploit the discoveries through an optical microscope to entertain us and improve our knowledge.

The Legacy of Tradition Hobbyist Microscopy
A list of entities, clubs, and resources still active,
along with addresses, web sites and a little about them
are in the appendices at the back of the book.

PART 2.
ENTRY INTO THE NEW 21ST CENTURY
HOBBYIST MICROSCOPY

Chapter 8. Contemporary Digital Photomicrography

Digital cameras have evolved over the last few years to match the image resolution of the older film cameras, although this is only really true of the higher priced SLRs (Single Lens reflex Cameras). The evolution hasn't stopped. Prices fall and specifications increase with each new release of the next range of models. It has never been easier to record the microscopic world photographically. Ideally, the SLR cameras can be fixed to the eye piece after removing the camera lens, using inexpensive T rings and adaptors.

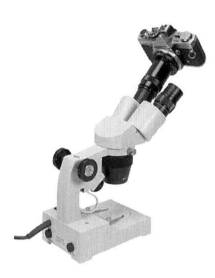

But lower priced compact digital still cameras have fixed lens systems which cannot be removed. Normally when taking standard photographs with a microscope, the lens system of the SLR is removed and only the camera back is used, this is because the microscope becomes the lens system of the camera. The majority of compact digital cameras and camcorders have lens systems that cannot be removed.

Fortunately other systems and methods including accessories have been developed for accommodating non-SLR cameras to record microscopic subjects under a microscope. **Brunel Microscopes Ltd.** (UK), for example, has developed its own universal non SLR digital camera adapter - the UNILINK

The Unilink will fit most cameras. Those that do not have an internal screw thread on the lens system will need an additional accessory - the LINKARM

The fact that the lens system cannot be removed makes it more difficult to attach the camera itself physically to the microscope than an

SLR. However some digital cameras - such as the Nikon Coolpix (990, 995, 4500) or the Olympus 4040, and most camcorders have a screw thread in the lens housing which can be used to directly attach an adapter via a coupling ring. Others such as the OlympusSP350 have a

screw thread used for adapters at the base of the telescopic lens. In general cameras that have a screw or bayonet thread related to the lens can be satisfactorily attached to the Unilink via step rings or adapters.

Almost all of the current models have telescopic lenses, and those with screw threads at the base require an adapter tube that allows space for the zoom before being attached to the Unilink. Unfortunately models change very rapidly and it is difficult to keep up with model design change, particularly as cameras are getting smaller and smaller. Those cameras/ camcorders without an integral screw thread can be attached using a device which uses the camera tripod bush to allow it to simply "look" down the eyepiece - the Linkarm.

Vibration
Unintended movement (camera shake) is death to good photography and the problem is accentuated when photographing objects hundreds of time smaller than those visible with the naked eye. You may notice how precariously balanced the arrangement of a camera mounted at the extremity of a long tube actually is, and realise this set up can quickly cause 'shake' even through the simple action of pressing a release button. Most SLR Cameras still have mechanisms to enable viewing through an eyepiece, where apparatus is 'clunked' out of the way during the moment the photograph is taken (shutter release). It is necessary in these cases to employ a shutter release cable to minimise vibration caused by touching the camera itself.

Depth of field
As with any other fields of photography, the laws governing light capture and definition come into play. The important one is what is called 'depth of field'. To simplify—a camera can focus sharply over a selection of the distance between the nearest point of interest and the furthest away towards an infinite point. The camera can not maintain

sharp focus over the entire range. As a rule of thumb, the smaller the hole (aperture) and longer focusing distances leads to a deeper depth of field and wider apertures and closer focusing distances lead to a shallower depth of field. A longer depth of field means more of the view passing from front to back will be in focus. A shallow depth of field means a thinner depth of the view will remain in sharp focus. See the diagram below.

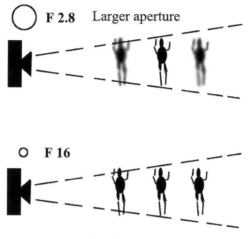

Smaller aperture

But of course, the smaller the aperture (hole) through which light passes, then the less light is passing through and thus the image grows exponentially dimmer. This means the shutter must be open longer to expose the digital sensor (ccd) to the light in order to obtain a satisfactory image. And this increases the chance for camera shake. The problem is far more profound when trying to photograph objects and specimens through a microscope. Whereas in the macro world we speak of focus lengths and depths of field ranging over metres, in the micro-world we are dealing with depths of fields and total focus length distances of just tens of micrometres (μm). To give you some idea of this very narrow distance:

1 to 10 μm: diameter of a typical bacterium
3 to 4 μm : size of a typical yeast cell
7 μm : diameter of human red blood cells
9 μm : thickness of the tape in a 120-minute compact cassette
10 μm : size of a fog, mist or cloud water droplet
170 μm : thickness of a glass cover slip

You can, for example, photograph an amoeba which has escaped a water droplet on a slide through spillage, and started to crawl across the surface of the cover slip, and not even see the reverse side of the cover slip because it has blurred out of existence.

When an SLR camera minus its lens is mounted onto a microscope, the microscope itself becomes the lens and the sub-stage iris aperture replaces the camera aperture and thus defines the F stop.

What does this all mean?

In one statement: it is *impossible* to photograph most microscopic subjects such that the whole specimen is in focus in the final picture! But thanks to modern digital technology, computer processing, and intelligently designed software—a solution is available. The solution is called Focus Stacking. (Also called 'image stacking').

Focus Stacking software

The idea behind stacking software is this: if you can focus and photograph a specimen at different focus depths from top through to the bottom of it, in each photograph some elements will be in sharp focus and some will not. By having software designed to realise the sharp focused elements in each photograph in the set, and extract that information—a new image/photograph can be created with all the focused elements combined into the single composite image.

Several software packages can be purchased to perform that very task. Some of the high end professional packages can even enable linking via stepping motors to the focus controls on the microscope, automating the entire process. But for us, trying to keep our expenditure down, and probably with budget priced microscopes lacking the finer focus ranging capability of more expensive and sophisticated instruments, we require an easy and cost-free solution. Enter our hero of the day, Alan Hadley, and his *Free* Focus Stacking software called Combine CZ.

The free photo-stacking software can be downloaded from here: **www.hadleyweb.pwp.blueyonder.co.uk**

The software achieves different results according to the complexity and type of specimen photographed. Of course, this can only be used where the specimen is static and not swimming or moving around. You arrange a point either at the very lowest or the very highest level and take a photo. Adjust the focus a very tiny amount to move continually in one direction, stopping and taking another photo after each tweak. Depending on how thick the specimen is, you will end up with from say, three to 25 photos.

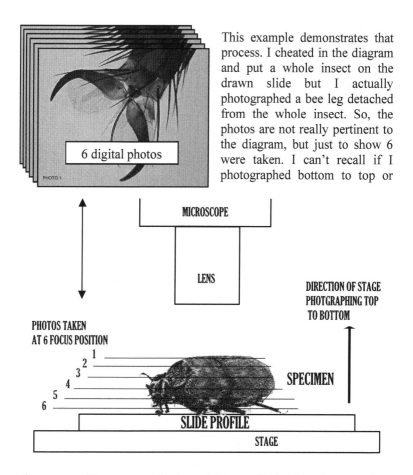

This example demonstrates that process. I cheated in the diagram and put a whole insect on the drawn slide but I actually photographed a bee leg detached from the whole insect. So, the photos are not really pertinent to the diagram, but just to show 6 were taken. I can't recall if I photographed bottom to top or

the reverse. Thus my numbering might actually be the other way than shown, but you'll get the idea.

The software allows for several different methods to be explored and some post editing work to improve the results. The specimen can be solid or transparent. The software can auto align each image such that if the specimen is not exactly in the same position in each photo, the software will ensure they are before doing its magic. A little practice will soon see you getting optimum results..

I used a very low budget microscope to record six images in my example of focus stacking. It was a bee claw. When you obtain a composite image of all the 6 photos and the elements focused well in each one now assembled in the final one, you may need to crop it.

My results are on the next few pages.

PHOTO 1

It's quite difficult to see the gradual difference in each photo but I have marked at least 1 area to look at for you to determine the difference in resolution and element separation in those areas.

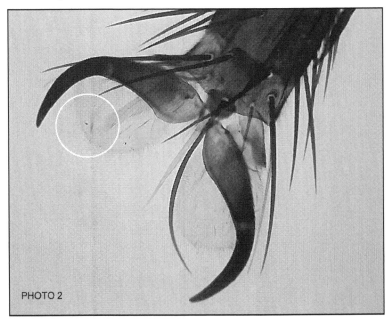

PHOTO 2

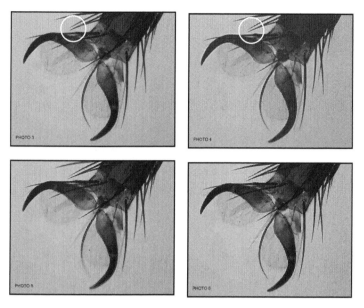

I have reduced the size of the 4 above just to conserve book space and you should understand the general principle by now. The final composite image is below. Compare this one to the 2 large ones on the previous page.

FINAL COMBINED IMAGE

You can also take a short video whilst turning the focus gradually up or down. The software will sample frames from your clip to produce a set of stills to combine.

When we study through the microscope directly, we constantly tweak the focus to enable our minds to build up information into a kind of holistic understanding of the specimen viewed. What focus-stacking offers as an advantage, at least on inert subjects, is a visual representation of what our minds actually do. Here magically, seemingly defying the rules of lenses, apertures, light levels and depths of field is a completely focused image. The Victorians would have been beside themselves to have such a clever tool and method to achieve this.

Focus stacking (some folks call it image stacking), can be used in Macro photography too.

Macro photography

I consider Macro-photography is still a microscopy-related pursuit simply because we are getting closer to small subjects and 'seeing' them in more magnified detail than we can with the naked eye. A macro lens can be used instead of a normal camera lens where a camera allows for different lens to attach to the camera back (most SLRs) or extension tubes can be exploited to move the existing lens forwards.

Non-SLR digital cameras may have a macro mode built in. If not, they might have an inner thread inside the fixed lens. Convertor lenses may be available which can be screwed onto that lens to allow a limited macro function. If neither of these options are applicable—one adaptor which is very versatile and of good quality is the *Raynox DCR-250 Macro Attachment available at the time of writing on Amazon for approx. £40.00.*

Raynox DCR-250 Macro Attachment

The clever thing about this adaptor is the way side-sprung clips allow for it to be quickly and easily attached to a range of cameras with no inbuilt macro facilities or cameras with no inner thread in the fixed lens. It can be used on SLRs too, saving the cost of dedicated macro lenses. I have used it on my canon 550D, for example.

SLR camera owners have far more options with the cheapest one being to use an adaptor ring to reverse fit an existing lens to their camera. Although this does disable all auto-functions in the camera relating to the lens as the connecting flat pins no longer make contact with the body electronics connectors.

I would advise that any person serious about macro-photography goes the distance and invests in the right camera and dedicated macro lenses to achieve more than average results. Macro-photography and Macro-video recording is easy to do badly, as you can readily witness on internet sites like youtube, but is far more involved to achieve good results. It requires patience, a good understanding of light levels, depth of field and knowing where the relevant controls are on your camera. Bees flying in and out of flower pollen hot-spots seem to have an uncanny sense of your proximity to them and skip off quickly when you *think* you are getting closer, undetected. Also... well.. they're busy bees: they're not going to hang around while you fumble with camera settings, instead of knowing them instinctively to keep a bee in focus on a flower head gently swaying in and out of focus. Even a slight breeze will ruin your best planned shot!

A well achieved macro photograph or movie can be a stunning vista to behold. Often, filming at faster fps (frames per second) and slowing down the footage afterwards can yield details lost when filming at 25 or 24 frames per second.. But more on that later.

Cheap options for photographing microscopic subjects.
If you are young, unless you have rich parents or family members, even the declining cost of good quality cameras make them resources out of your reach. We are all stuck with having to live and achieve things within our limited resources, but it need not stop you achieving the best results for your tools and equipment. And there are tricks... as I will explain.

Lower cost cameras can simply be 'plugged' into a tube in place of an eyepiece to record videos of photos to a computer. Low budgets means low resolution cameras. But, resourceful and imaginative people can often find work-arounds for the limitations imposed on them. It requires learning more profound details about a given thing rather than accepting quick tips and generalised statements proliferated by commercial web sites and physical popular science magazines. But this is where the breed of amateur microscopists share a set of common traits: attention to detail, passion, intelligence, and the capacity to look further than most of the people around them.

Cameras, sensors, megapixels, and imaging

If you have an expensive camera with, say... a 15 megapixel camera, and another one with, say... a 2 megapixel camera, and if everything else was equal—the quality of the optics (lens), the mechanism for focusing and setting aperture, plus the microscope you will use to record an image from—which camera do you think will record the most resolved detail?

In reality, it is likely the 15 megapixel sensor will be contained in a higher specified camera and the lower megapixel sensor in a lesser one, but we can ignore this for a moment.

Before digital sensors (charge-coupled devices) were used for photography, a film coated with an emulsion was used to record light by chemical changes to the tiny molecules of the emulsion. The more grains of light-affected chemical you could squeeze into an area of film, the more resolution and detail you could potentially record.

Just suppose for one moment you have a digital camera today with a one pixel sensor (not 1 megapixel), just a single pixel but you need to use that camera to record an important moment, visually. Let's make the problem harder still. Maybe that 1 pixel sensor can't even distinguish colour. Instead it can just distinguish black and white. But the task before you is to create (record) an image which is thousands of pixels wide and as many in height with fine detail in full colour of the subject you wish to record an image of. Can it be done?

Amazingly, yes it can.

And the same solution can enable lower resolution cameras to achieve results equal to higher resolution cameras (hypothetically). In practice, it won't—but the method does provide a way to achieve enhanced results with low specification gear.

The reason why this is possible is because a microscope is being used to transmit light containing image information to the recording system.

Large image capture. Photo-stitching: a demonstration.

The following original image, if printed out onto paper, would measure 3 feet by 2 feet at a resolution of 300 dots per inch: the required DPI minimum for 'respectable' minimum print resolution. It was created using a mid-range camera—not one with a sensor considered to be a professionally-recognised one for publication quality images, and by using it on a microscope affordable by young people—a monocular

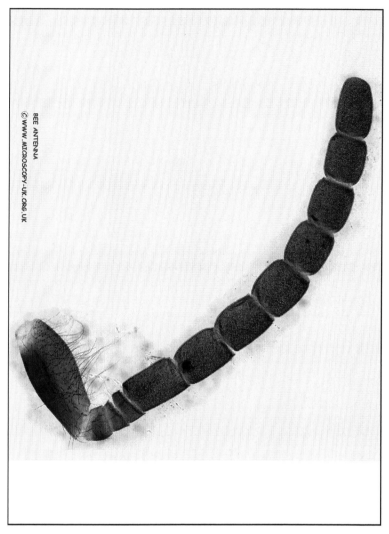

costing less than £80.00. It is greatly reduced in size here, of course.

The secret in producing this image is that it was not a single photograph. The camera was attached to the microscope, the slide was arranged such that the top left hand corner of the specimen was in view, and then the photograph was taken. The stage is then moved in one direction horizontally (X axis), and then another photograph is taken, moved again, another etc. When the extreme right edge has been 'snapped' the stage is then moved vertically—the 'Y' axis, and the process continues again in the opposite direction. The panning and photographing continues until the entire specimen has been captured on camera, a section at a time. One has to judge when panning, how much to move the stage. It is important to over-lap each photograph because once we have a complete set of images, special software will 'stitch' the images into a whole larger picture. The software needs clues as to where each image belongs and where to position it.

The software is not perfect and you have to be careful to keep the light levels even over the field of view and in the same level in each shot. A degree of post processing may be required to tidy the final image up.

I found the best photo-stitching software was already included with Adobe Premiere CS4 extended under the menu item 'File', Automate, Photomerge.

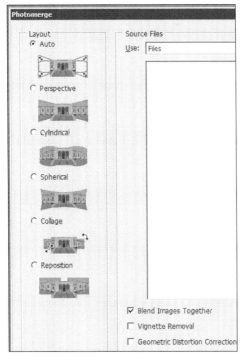

Try Auto first. If you haven't got a Photoshop version with this facility, there are stand-alone software packages which do similar.

I haven't tested any other others, and there are quite a few, but you can explore the options they have to offer by visiting a page on the internet which lists them all and displays the key features each one has (*see below*).

If you have the patience, you can take several photos of the slide in each position and photo-stack them to improve focus. You would use the output images to photostitch together.

A list of photostitching software:
http://en.wikipedia.org/wiki/comparison_of_photo_stitching_software

Putting large images on the internet
Displaying your images on the internet or sharing with others is made more difficult when producing very large images. Although the dots per inch (DPI) are more commonly 72 or 90 dpi on a digital display

rather than the 300 dpi required for prints, a 36 inch by 24 inch is going to be absolutely enormous. This is where a nifty bit of software comes into play called: **Zoomify**
This is what the makers of Zoomify say about their software:

Zoomify unleashes unlimited resolution.
Zoomify makes high-quality images useful on the Web. The Zoomify family of products quickly converts images of any size or quality to stream for fast initial display and on demand viewing of fine details!
Zoomify enables publishing of multi-megabyte or even multi-gigabyte photos that can beviewed without any large download. Visitors to your website can interactively zoom-in and explore huge images – of truly high quality – in real-time.
Zoomify's solutions are geared toward the needs of your website.
Zoomify HTML5 enables your site visitors to interactively pan and zoom to explore images in detail with only a web browser. Zoomify HTML5 supports web page design flexibility with simple HTML parameters.
Zoomify HTML5 Developer adds extensive additional features including layout control, copyright alerting, watermarks, and more –
all driven by HTML parameters, along with helpful examples, as well as full editable JavaScript source for the Zoomify Image Viewer for complete customization control.
With no Flash or other proprietary technology involved, your Zoomify content works everywhere – even on iOS, Android, and other mobile devices.

What it does is cut up a large image into many small rectangles and then writes the code to enable an html-styled web page to combine those rectangles into a series of larger view points under the control of the user. I used this software on my Microscopy-UK web site to build a 2D microscope. It is a software suite absolutely perfect for the microscopist to display his/her images on the World Wide Web, or indeed—on digital displays in schools, museums and other learning institutes. Two screen shots of this facility on my web site are opposite. The first picture shows the whole specimen in the slide. The user can then zoom in repeatedly a number of times and pan each zoomed image.
Fortunately, I'm a computer programmer too... or at least I was professionally. These days, I just dabble with coding. Anyone with a rudimentary understanding of the Javascript computer language can...

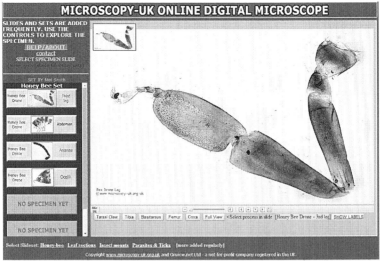

Screen shots of: my 2D Microscope which uses photo-stitching software to source the image and zoomify to present it in a zoom-able and panning fashion on the internet.

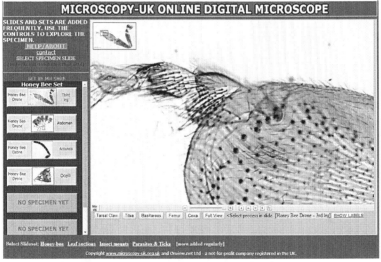

(...continued from previous page)
...do what I did here—namely to create a snazzy interface so visual virtual specimen slides, to the left in each image) can be loaded into the zoomify viewer. The best way to see this is on the internet at:

http://www.microscopy-uk.org.uk/2D-microscope/

The three images on the right are from the several small tiles from over 1500 generated by zoomify to display the large one in the novel way it does.

A slight pause

Just to explain a little bit about my reasons for introducing post-image processing and display guidance in this book, I believe we are now living in a far different world than all people who practiced hobbyist microscopy in the past. If they had wanted to share images of their studies, their route would have initially been through drawn illustrations like Robert Hooke did in his book—Micrographia, and later through photography. Few journals existed to publish their images.

The contemporary microscopist today has so many avenues he/she can choose to share their work with others. It is important therefore to assist microscopists in achieving that sharing in novel and often superior ways. As an aside (and knowing that many enthusiast microscopists dislike comparison with this other amateur pursuit) astronomy appears much better presented on the internet than many amateur microscopy web sites.

I believe if microscopy, as an enthusiast pursuit, is going to survive and flourish, then presentation to

others of what we see with our instruments should be at a level which is engaging, informative, and intelligently novel.

I also believe contemporary and evolving technologies are very apt for the field of microscopy, its study, and its presentation. We will explore all these here in the following chapters, beginning with stereoscopic 3D microscopy.

Chapter 9. 3D Microscopy — An introduction

Prologue

The term 3D when applied to imagery has become ambiguous. Ten years or more ago, most people would associate 3D imagery with those pictures that have blue/red or green red fringing around elements in the image (anaglyphs)—intended for viewing with appropriately coloured glasses, or with the cards which have two almost identical images printed on them and viewed through a two lens viewing system. Indeed, these types of 3D presentation still survive.

But now, as you may have witnessed in 3D cinemas, we also have 3D imagery where a left eye/right eye image pair is combined and projected through a polarised light system and viewed using polarised spectacles or glasses. Digital displays, televisions, blue ray discs, computer games, and even channel TV all now have adopted limited 3D viewing experiences.

And then there is lenticular 3D imagery where two or more images are broken down into a series of fine lines and subsequently viewed through a flat, transparent, lenticular plastic sheet, or on a digital screen employing a lenticular mask.

But an additional technology has also come barging into the 3D imagery family which is distinctly different to all the others: 3D computer modelling. I will leave out the technology of 3D achieved through light holography (where light interference patterns and lasers are involved in image creation—a hologram), because the technology is still beyond ordinary peoples' reach due to the costs involved and the equipment used.

As I write this book, 3D printing is becoming an affordable technology for the home consumer market as prices for the printers are now within the budgets of non-industrial application. The new amateur microscopist of this era is presented with an array of remarkable ways to expand their hobby through grasping and exploiting all of these 3D technologies. I'll expand briefly here on each of the ways of producing and showing imagery in 3D, in case some of the readers are not familiar with them. The following chapters will expand on this one and guide you step by step into how to produce 3D images from your microscope. These techniques can be applied to both still and moving images (video).

Two eyes

Our brain receives visual information regarding the external world from our two eyes which see out from two different positions. They may only be a tiny distance apart but it's enough to ensure objects, especially ones closest to us, are perceived from two different angles. Each view of the 3-dimension eternal view contains slight different information. This, along with size and in some instances sharpness and brightness, is interpreted and cross-referred by integral mind processes such that you see a single combined view of the world. And... it's in 3D.

The trouble now, of course, is that when you look at a photograph or the things in it, as it has been recorded from just one position and viewpoint, it lacks any information regarding depth in the way our brains decode it. Yes, size of objects relative to others in the image, dimming and less sharpness of objects further away to the position of the device recording the image... these things give us clues about distance but they still do not give us what our brains need to perceive an image in 3D. The mind needs two viewpoints, not one! And those two viewpoints must equate approximately to the distance between the two eyes of an average person.

Stereo image pairs—cross-eyed.
You can actually look at two different pictures, photos of the same scene or object, and provided they have been photographed at the rough distance apart as two human eyes, you can use a method to cheat your eyes so that each eye only sees one picture and not two. When you

do that, the image magically becomes a 3D one. Look at the double image below, but *cross-eyed.*

You might need to go closer or further away to cross your eyes without too much strain. As the pictures come together to form a third one in the middle of the two... concentrate on that one. It will become clearer in 3D the longer your mind accommodates the 3D clues in the two pictures. In this example, the image shown left above was recorded from a left eye point of view and the one on the left from a right eye point of view.

It's just easier for most people to go cross-eyed, rather than parallel-eyed. In the later case we would display the pair of images the other way round... left-right. In that case, we could use a pair of lenses to relay the correct image to the correct eye and you would see the picture in 3D without needing to look silly by crossing your eyes.

Here is the type of basic 3D viewer *(right)* you may be familiar with. The mars landscape (courtesy of NASA) is a pair of images where left view is left and right view is right. I have removed it and placed it below, with left/right reversed, so you can look at it crossed eyed.

Mars Image (c) NASA but placed in the public domain

You need to get quite close as the image here is very small.

Anaglyph

There is another way to get the information in the left image to the left eye and the information in the right image to the right eye without using this type of viewer. Unfortunately it uses colour and this is a black and white book. The image (left) if you look very, very closely appears blurred. Normally, in a colour book, you would see the black and white details have fringes of red and cyan. This is because the cyan colour channel of the left hand image in a stereo pair is removed and replaced with the cyan channel of the right hand image. The combined image can be separated by putting on red/cyan filter glasses. The two coloured lenses 'trick' your eyes into receiving the information—red in one eye, cyan in the other—and your brain provides the processing to build the 3D representation.

Polarised light

In a modern 3D movie at the cinema, a digital projector employs polarised light to transmit and overlay the left right images on the screen. Glasses containing polarised lenses either at 45 degrees or 135 degrees difference through the use of the polarising material, filters the light reflected from the screen to the appropriate eye. [*Note: two different systems are used. See diagrams below*]. There is also a system called shutter glasses 3D (now mostly out of date since the technology is more expensive and cumbersome) where the left image is displayed for a tiny fraction of a second and then the right one. This repeats rapidly. In this instance the shutter glasses use lenses filled with electrically polarising crystals which are in synchronisation with the rapidly flickering screen. Vision through one eye is obscured and then the other, left, right, left, right in time with the image display. The mind is fooled into receiving left and right image information so rapidly that it doesn't realise there is any delay between the two images and interprets the screen as a flicker-less 3D image.

Linear Polarisation of light

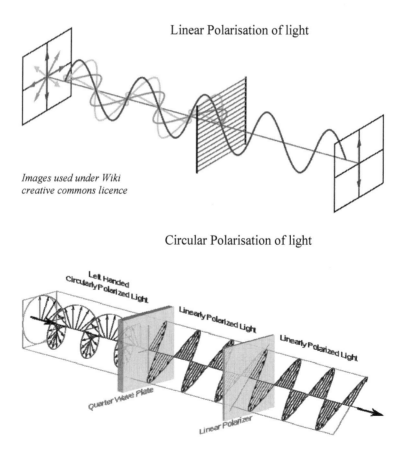

Images used under Wiki creative commons licence

Circular Polarisation of light

The stereo microscope
It doesn't take much of a leap to realise a stereo binocular microscope is a perfect instrument for recording left/right image pairs or left/right movies. And the usefulness is greatly increased since most things observed through a stereo microscope are unlikely to be veering towards one dimensional objects, which is often the case viewing tissues for example under a compound monocular microscope.

If we wish to just take 3D photographs of inert and non-living subjects, we only need one camera, but if we want to make 3D movies where the specimen is a live insect for example, and active—we need two cameras. The latter need not be expensive but you might need two computers to manage the recording of two USB digital cameras to your hard disk. Most home-consumer computers would struggle to accept two such inputs.

But with one camera it's simply a matter of photographing the subject through one eyepiece tube and then moving the camera to the other one and taking a second photo. A little bit of fussing around to get the camera orientation correct and some post work to ensure the views are aligned will do the job.

The big internet presentation problem
Good. So we have a means to record 3D images and films of microscopic and macroscopic forms, but now comes the problem. Most computer users do not have 3D displays and most of them probably do not own red/cyan glasses. So, if we wish to share our microscopy 3D wonders, how do we do it?

A solution exists and it is a by product of this other 3D image display method which I have not explained yet. In fact, it is so important to our aims that it deserves a complete chapter of its own. This method opens up the opportunity to share our 3D microscopy images world-wide as hopefully, you will see in the next chapter.

Chapter 10. Lenticular Imaging

The technology has been around for decades but in the past it has been used as a novelty rather than a refined way of showing both moving and 3D images in a single seemingly flat plane. You may already have seen lenticular images on DVD covers, posters, and advertising boards.

Many older people may remember the fuzzy screens on kids toys which animated when you wiggled them. Well this old technology is back with a vengeance and advertising posters are starting to appear in public areas which move, flip, or project in real 3D.

Lenticular 3D images rely on a plastic micro lens system, normally thin sheets, stuck over a composite image. The image itself is an interlaced picture. The interlacing say, of a stereo pair needs to be processed first to create a series of images as though they had been captured from a series of viewing angle and not the original two.

The composite image would consist of vertical strips from each image alternating left to right across the composite picture. The strips are very thin and each one is designed to be projected out by a linear lens (just that one strip), but there can be hundreds of strips and dedicated linear lenses per horizontal inch. All the set of strips say, from the left picture will be projected by its appropriately aligned lens to the left eye and all the alternate strips from the right picture project to the right eye. As you swivel the picture or you move from side to side, another set of strips would do the same from the next view point. Depending on how many strips per inch there are, has a bearing on how much you might see thin vertical lines in the image.

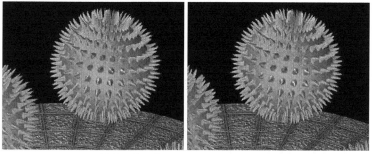

Left and right view image pair

Let's take two photographs of a geometrical shape. It could actually be a microscopic form like Volvox but to keep it simple, I'll use a spiky ball (*see above*).

You can check these two would make a stereo pair as I have reversed the images (left-to-right) so you can see a 3D representation through cross-eyed viewing. They do not look that different but they have been captured from two different positions so if you overlaid one on the other, the differences will show up and a blurred image will result. I have overlaid them here (*right*) so you can see they are indeed, different. There are software packages available

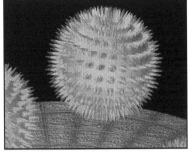

Spiky ball image created by Brian Johnston—a leading contributor to Micscape

which will now create a lenticular-ready composite from the two images (the left and right ones). I use several software programs in my 3D work but to create a good lenticular image which looks 3D, you need software to process the stereo pair to create a series of images. I use **Triaxes StereoTracer.** The program achieves a very clever thing.

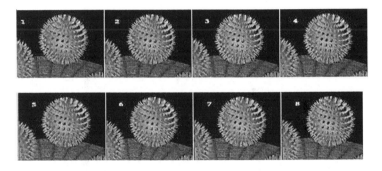

It makes in-between images. Each one is as though it were a photograph taken from a position between the original left and right positions. See diagram right. Only the images captured of the object from camera L and camera R are real. These images are the stereo pair. The software creates images as though the virtual cameras and view points were real. In a lenticular picture, if we just used the left and right image the resultant effect would be a 'flip' image and not the

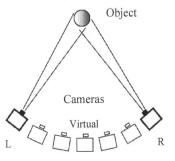

3D one we desire.

A 'flip' image would cause the image to change to a second image as you twisted it back and forth in your hand or your eyes moved left to right across it.

Once the software has produced the in-between views, we export them for use in any of a number of software packages which will construct a composite lenticular image by taking thin strips from each of the views and combining them according to the pitch angle and lenses per inch of the lenticular screen to be used to view it.

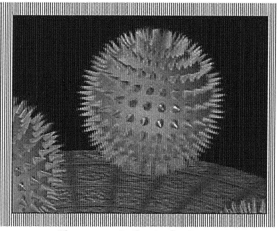

The composite lenticular image.

A software package I've used to do that is ***3D Master Kit*** but a

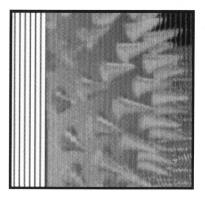

Detail from the image above.

search on the Internet will reveal more and they'll probably be cheaper to buy than my one.

Now, that is all well and good, and you could print out the image on a good printer, buy the appropriate screen, align the image and stick the image to the screen. But let me warn you: *I've done that and it is very hard to do!* Far better to send it off to a company who will take your composite picture and mount it on the screen for you.

But... now here is the biggie! To

distribute our 3D image across the internet, we do *not* want to produce a lenticular print. You can do that if you wish and it will look great but we are interested here in sharing our 3D images with other amateur microscopists and members of the public who might like them too.

What we want is those in-between views created by Triaxes **StereoTracer.** This piece of software (and there are other similar software packages) can create an animated *.gif* file or a movie from the series of images. These can be inserted into a web page and the viewer will see the 3D object swivel.

Examples are here on my Microscopy web site at:

http://www.microscopy-uk.org.uk/3d-microscope/index.html

Even more clever still
Something far cleverer is that you can actually create a 3D image of a microscopic subject without have a stereo image pair in the first place. You just need one original image! Albeit—purists will say this is a cheat, but my view is this: microscopy is already an area where an optical microscope helps one perceive processes and details in an object or an organic form which is difficult to see and difficult to interpret. Many of the specimens have very little depth in the first place and a slice of tissue is really a 2D slice through a 3D entity. It's a bit like looking at ariel photographs and working out what are really building, cows, missiles(?). If 3D is how our minds normally perceive things in our macro world, then 3D is part of our normal process and mechanism of interpreting whatever it is we see or study.

StereoTracer (and similar software programs) can exploit something called a depth map to help produce a set of images of a scene/object or it can create a 2 image stereo pair, just using the one real photograph and a depth map. So, what is a depth map exactly?

A gray-scale depth map image is one where a duplicate of the original photo is shaded in with levels of grey ranging in tone from brilliant white to deep black. Where the grey is whitest, that element of the image is deemed closest to the viewer. Where the grey is darkest, that part is furthest away. The best way to create a grey depth-map image is manually, but Triaxes StereoTracer has a built in function to generate an automatic depth map... not too successfully though, in my opinion.

The best way to create a depth map image is manually in a photo-editing software package such as Adobe Premiere. It takes a lot of patience and quite a bit of skill, but the rewards are spectacular. The more you create them, the easier it gets and as a by product—it actually helps you study the physical structure of any given specimen. Allow me to demonstrate on something simple first. Remember the original image can be a colour one or a black & white one.

A simple shaded cube

To create a depth map of this cube, we duplicate it and shade it thus...

The depth map image

The edge of the cube closest to you, I've made whitest. Each face grows a darker shade of grey as it recedes away from you. The background is deemed infinitely distant and I coloured it black. It is best to blur the depth map image a little (I use Gaussian blur) as the software works best on an image without sharp edges in the details of the depth map.

The depth map image—now blurred!

If you now use the StereoTracer program it will generate a swivel .gif image or a movie of the cube appearing to swivel in 3D. The principles here can be applied to microscopic subjects like this Diatom frustule shown here (*below*), greatly magnified.

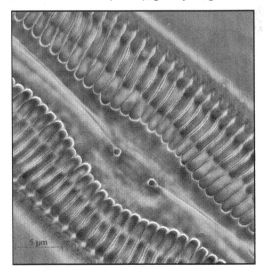

It's a more painstaking job creating the depth map images as you can see below. I had to carefully digitally air brush the fine projections whiter and the dips between them darker. You really have to see this this as a swivel 3D to appreciate it.

Take a look here at:

http://www.microscopy-uk.org.uk/3d-microscope/14/

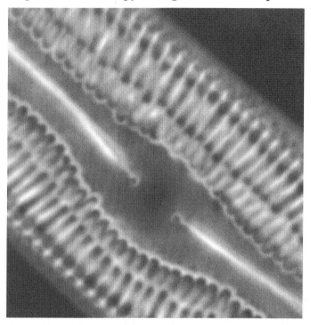

Greyscale depth map image created manually of the diatom

The fantastic thing is that once you have the depth map and modified it to produce a good effect (it might take several goes to get right), you can use software to make anaglyph 3Ds, cross-eyed stereo pairs, lenticular images, and swivel files as Animated Gif or in movie format files.

Let me show you the power of this. So, you've seen the diatom as photographed through a microscope on the previous page. Only one photo was taken, yet miraculously—you can now see it right here on the other page in 3D without any aids and without this being a coloured book. The only better way I could show you this diatom frustule is by putting it under a good quality stereo microscope and letting you look down the tubes.

In fact, there's an idea. Maybe I should create a 3D Cross-eyed Microscopy Gallery book?

To see this image best, turn the book around and stare cross-eyed at the pair of images. Experiment with distances from your eyes until it feels right. The two images will merge to create a third one between them: a 3D diatom frustule. The longer you concentrate on the phantom middle one, the clearer and more stable it becomes.

This way up

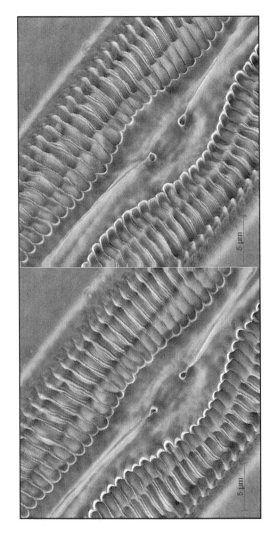

3D Diatom Frustule. View crossed eye by turning the book around.

List of resources for 3D, Holograms, and Lenticular Images

http://www.3dz.co.uk/
Will create 10"x8" prints of your lenticular images. They can create holograms too and have great information about the subject. UK based.

http://lenticular3d.com/
An online list of resources, groups, and good information.

Software
http://triaxes.com/
StereoTracer Software
3DMaster Kit Software
They also sell other related software. I use their software a lot as it's easy to use and reliable.

http://www.stereo3d.com/3dhome.htm
Alternative programs and software for stereo, 3D, and lenticular imager creation.

http://www.stereo.jpn.org/eng/
These guys have been going for years. Their **software is FREE** and I love everything they make. Visit this one first and get engaged.

http://www.microscopy-uk.org.uk/3d-microscope/
An online 3D virtual microscope. I did these with the help of contributors to *www.microscopty-uk.org.uk* and Micscape Magazine. Includes the fantastic imagery created by Dennis Kunkel using a SEM microscope in Hawaii.

http://www.ononesoftware.com/products/resize8/
Often, the images you create are too small to blow up to larger formats (remembering 300 dpi is the lowest print resolution)., but there is a 'fractal' method for blowing up images to huge sizes. I use software called *Genuine Fractals*, now called *Perfect Resize*, which performs the job brilliantly. Many professional printing companies use this software too. It's not cheap but it's very effective and one of the few solutions to this issue I've found. Also works as a plug-in for Adobe Premiere.

Chapter 11. 3D Computer Generated Models

You've seen them in the new generation of CGI movies and on high quality TV nature shows. If you play computer games, those cars, planes, spaceships are all from the same 21st century science: computer generated modelling. The term is not entirely correct. One or many people have to design the models using computer software to generate them.

The creation of 3D models by computer technology has permeated almost every human endeavour except... yup... you guessed it: microscopy!

There are 3D models of insects though, and lately I've seen a greater surge into models of true microscopic forms. Once again, many purists will be banging their heads onto their microscopes at my suggestion that 3D computer modelling should form part of the toolkit and study area of an amateur microscopist. They may well be right, but I would strongly disagree, not least for my belief that any young person interested in computer technology (and most are) who also becomes a keen and accurate amateur microscopist could easily become a multi-millionaire by tapping into these two areas of interest. I'll explain as I go. But first, an image presentation of one or two 3D generated models, and it's a shame I kept costs down, and you can't see these in colour...

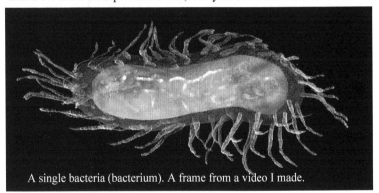

A single bacteria (bacterium). A frame from a video I made.

A frame from a video where the bacteria is spun around and animated so you can see the flagellum (tiny hairs) undulating, along with internal mechanisms and processes shining through from inside. You can't do this with an optical microscope and yes—it relies on the skill of the person creating the model to get the details accurate. And what that requires is someone well versed in microscopy, but in this case Scanning Electron Microscopy.

The more young people who realise the world is built bottom up and hobby in microscopy when they are young, the greater chance they will take with them powerful information to help them in the human struggle to eat, beat diseases, cancer, and sickness... and to discover new ways to manipulate our environment in ways beneficial to it, and the life forms which inhabit it.

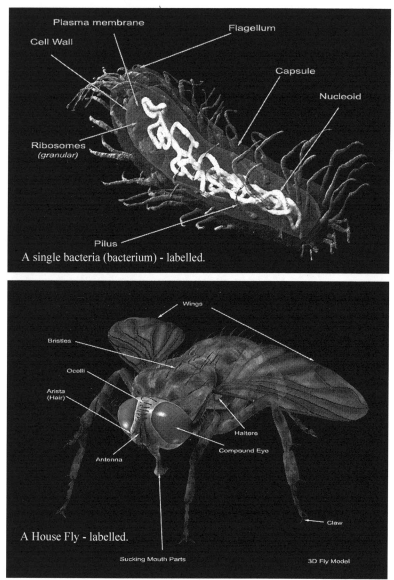

A single bacteria (bacterium) - labelled.

A House Fly - labelled.

For anyone who has never been involved with generating or using Computer Generated 3D models, I must explain a few elementary things. And since this subject could fill volumes of books. My guide is merely stroking the surface.

Creating a model
This is the toughest bit. Software is used by a designer to create the sculpture from basic geometrical shapes initially...

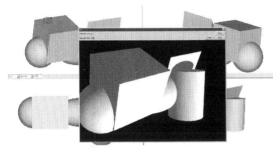

These shapes and structure are formed of tiny straight-edged geometrical blocks, pyramids or/and polygons. A final completed 3D object's quality is often partly determined by the number of small elements which make up the complete model. There is a trade-off going on regarding the computer processing capability to mathematically keep track of the thousands of co-ordinates (spatially-represented) of any given 3D form. The more discrete components there are, the longer the processing time to reproduce or create a render of the model.

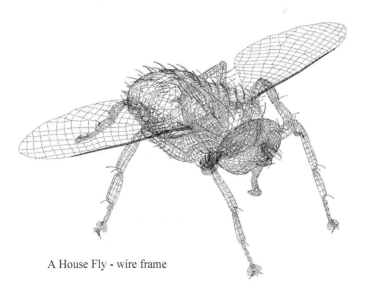

A House Fly - wire frame

Designing a model of this type cannot be animated unless it is also designed to be so. This involves the use of internal splines (sometimes called 'bones') which determine how the various sub-structures will move and react together.

Textures
A texture will be created which determines the colours of any surface or part of one, and the way light will react with the surface of the model. The extent of the surface 3D protrusion is determined by depth maps and other images such that a rendering package (software to interpret the model into a final image) can use them to create the illusion of non uniform surfaces.

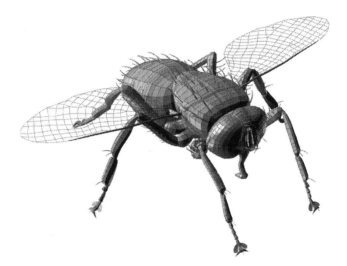

Both texture creation and the model itself and the way it articulates requires complete understanding of several different, often unrelated fields. Consider anatomical modelling compared to military vehicle modelling, for example.

Animation & Posing
To mimic or represent real things, computer models must be capable of behaving, moving, and also reacting with other models. Like a puppeteer, an animator can create sets of animations which are recorded and can be applied to the model later. This may include, for example in human based models, gestures, expressions, emotions etc. Even this resource can only be utilised successfully by the end user

who may use the models, their poses and prescribed animation routines, along with lighting systems (themselves created by software lights), shadow systems, to create a final movie, diagram, or other output resembling something which looks like it was a snapshot of our real external world or an entity within it.

Connections

My description is basic and rudimentary but I hope provides a general idea. I have spent many years exploring several elements of this subject and indeed, created both simple models and complex animation routines for models designed by others. The process of creating human based models and their movements has evolved greatly by using real people and indicators on them to record and track their movement (via the key indicators) and then map the result onto the 3D generated models. It stands to reason, this is not possible with living forms too small to attach markers to. Thus... the creation, movement and behaviour of microscopic entities and many macro ones which are extremely small (insects) are best recreated by microscopists or others who can study with an optical microscope. The textures and structures of tiny forms can only be created by the same.

Enhanced study of microscopic entities

Computer science and its application continually opens up new ways of understanding our reality. By mapping what we see onto computer generated models and by applying algorithms created using real live data, we are able to model 'what-if' scenarios - technology proven in aircraft design, space flight, and understanding the spread of viral and bacterial infections.

This advance will not stop. The ability to predict and anticipate outcomes within a complex reality is one of humankind's greatest assets. A contemporary young amateur microscopist is likely to be curious, intelligent, and resourceful. The idea of him/her using true study of real things and life forms by the use of a microscope and the comprehension of 3D modelling associated with it can lead to significant outcomes for the realm of amateur microscopy, and in a professional environment, to considerable advantages and discoveries.

Beginning

3D modelling and manipulation software prices ranges from the cheap to the 'what bank shall we rob' to afford it. One of the longest standing entry software suites is called **Poser.** And it's probably the easiest to use to create images from existing models, which are purchased online. A contender to this is **DazStudio**, with many fresh entrants to

3D posing and image creation preferring it. Note that these two packages are more about making final images by loading and posing or animating models designed and created by others. If you wish to go down the route of creating your own models, many packages exist to do that but costs rise steeply for the most effective ones.

The cost of models to licence varies depending on how complex they are and how complete they are in their build. The Vegetable Eukaryote Cell (above) would cost $149.00 to buy, but you can then

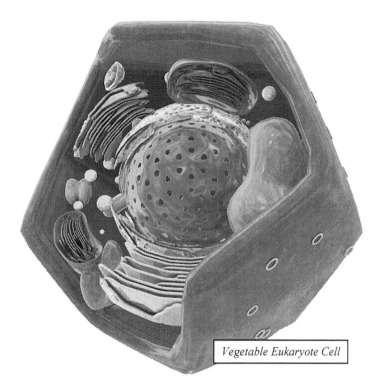

Vegetable Eukaryote Cell

use your model in any kind of output except selling the actual model on. Any budding teenager (or mature person) with the right skills to create 3D models would make a good financial return today if he/she chooses areas which are in demand.

I mentioned earlier that the 3D models use texture maps. The cell above, for example looks pretty 'naked' without its various surfaces correctly shaded and textured—as you can see in the image on the next page.

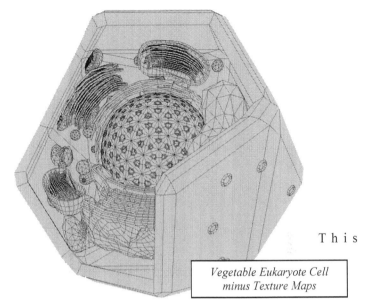

Vegetable Eukaryote Cell
minus Texture Maps

T h i s

cell, and many like them can be licensed from a long running depository for good quality models called Turbosquid at *http:// www.turbosquid.com/*

The fly on a previous page has a texture map too (*lower left*) in full colour for the original

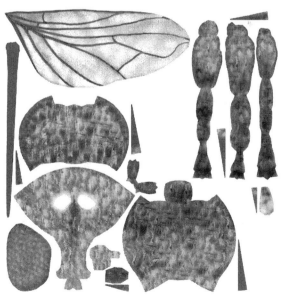

You can see how there are textures for each part of the fly including its body hairs. Each sub-part of this c o m p o u n d texture image map must wrap s e a m l e s s l y around the appropriate sculptured body part.

Below is the working screen for Poser which allows loading, image-mapping, lighting and animating 3D models.

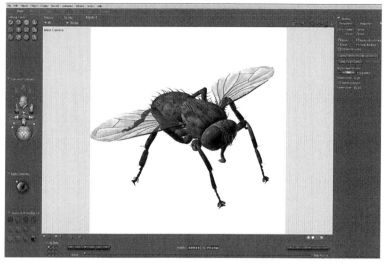

An entire HD tutorial video can be made by spinning the model, getting in close to its processes over a series of frames and then exporting them as stills or a ready made video. Labels may be put in and voice-overs afterwards by using a video-editing suite like Adobe Premiere.

Resources
Posing and animation of models
Poser: http://poser.smithmicro.com/
DazStudio: https://www.daz3d.com/get_studio   (free versions).

Model creation software
Comprehensive list at:
http://3d.about.com/od/A-Guide-To-3D-Software/tp/Full-3D-Suites.htm

Microscopy-related 3D models in action on my web site:
About them:
http://www.microscopy-uk.org.uk/mag/artjan10/mol1.html

Microscopy-relate 3D models I've used:
http://www.microscopy-uk.org.uk/3dmodels-index.html

Chapter 12. Making Microscopy Movies

I remember a time when there were no videos on the internet, now it is full of them, and most of them are poorly made. The worst being the many microscopy related videos on youtube. People, thinking they are now film-makers, stick a camera on a microscope, set it running and record the least inspiring and least informed tripe possible. Yes. There are exceptions. But they are few.

If one wanted to make a documentary film say, about people going to work in the morning in a city, sticking a video camera pointed at a railway or underground station entrance and turning it on for 30 minutes does not a classic make. What it does do however is transfer utter boredom to the viewer and probably an immediate rejection of anything to do with that topic.

I make full length feature films, short narrative movies and the odd microscopy documentary. Making a 'proper' film requires many skills and a desire to tell a story well. There is no reason to think the same skills required to make a narrative or documentary film about things in the normal world do not apply to the drama under the microscope, because it does. And just like a normal film say, about travel, requires a good camera operator, a film with content recorded at an optical microscope needs a good amateur microscopist.

David Attenborough is renowned and respected for his Nature films. Go watch one and see how the structure educates and informs at the same time. He uses drama too, often depicting the competitive nature of living forms to survive, hunt, avoid predation, and reproduce in a world where all resources have to be fought for and won. Any amateur microscopist today making a good movie for either the web or broadcast TV—and now we all can as HD cameras are cheap—will stand out from the crowd and make a name for themselves relatively quickly.

How to make a great microscopy or macro movie—introduction
The first thing is to decide what topic or subject you think you might make a film about. You don't need to think of a story yet, just the subject. Let's assume you decide to make a film about Desmids, or Rotifers, will it be a generally aimed documentary—one which just introduces rotifers, say—but not go into a lot of detail. Or, do you wish to make a film about a specific type or types of rotifers?

A topic which might appeal to a maximum audience would be the microscopic world of pond-life, but that alone would be a vast subject to cover in a single film, even if the film were an hour in length.

The next problem you have is in realising that whatever subject you wish to cover, you are going to have to get specimens to film, and you are not going to know in advance how they are going to behave when you film them. Rotifers eat, get eaten, give birth, interact with other life forms (in basic ways). You won't know in advance what you are going to be lucky enough to record!

What other resources might you use to 'liven-up' your film? There is sound, music, sound effects, and dialogue—all important and fairly obvious. You can use on-screen labels, split screen content, 3D model animations, a human being talking to the camera... all elements which will help make your film stand out from the rest.

I've been making a series of films to introduce young people (and new adults) to the wonders of amateur microscopy. *You can watch them free online at:*

http://www.pippasprogress.net

Learning by example

The best way to show you how to make a good microscopy-related video is by working through one of mine. This may help you realise the issues involved and the best ways of solving them. My video series employs the young lady Pippa to help create interest, and my videos introduce ideas and thoughts (apparently hers) not directly related to microscopy study per se. Your videos may be more about what's under the microscope, but the same considerations I had should be what you should also bear in mind. I will discuss the hardware side—cameras, sound recording, editing as I go.

Right then, maybe it would be best to get online and watch this episode of Pippa's Progress: **PIPPA'S PROGRESS EPISODE 3.** You will be able to select it from the web address above. It lasts fifteen minutes or so.

Before you start making your video—planning

The first thing to learn is this: all the video clips she shows you under the microscope, I already filmed before Pippa arrived to make the film. What I did is went out and took specimens from streams and ponds, and after looking at what I had brought back, filmed some of them. It was not summer or warm when I did this so it wasn't the best time to collect freshwater specimens. In fact, what I found I had collected was not the best choice. However, this video was not really intended to go into a lot of detail. I hoped to aim it at instilling enthusiasm in the viewer to have a go at what my star Pippa does.

This is important. You need to try and understand what your objectives are: just entertainment, pure education, a mix, a 'wow' factor, and... is it drama, informative, can it be edited together properly so all of the parts will form a whole to achieve your objectives? In this episode, these were my **starting objectives**.

1) Ensure Pippa is central to the film so that other young people are encouraged to emulate her.
2) Show that freshwater life microscopy can involve pleasant days out in pleasing environments.
3) Show you don't need top notch equipment and if things go wrong, you can improvise.
4) Make sure it's not just a jolly day out. Show tthat she planned well and brought things with her, and she's adhering to basic scientific recording principles.
5) Ensure safety is maintained because people (parents) are over-sensitive to that despite the real minimal risks involved.

It's important for you to see how I establish these in the first *one and a half* minutes:

Objective 1:

Stills of Pippa first, with appropriate music to set the mood.

Objective 2:

Cut to a video clip of her in that nice place—by a stream.

Objective 3:

Show she doesn't own a proper net but she is going to improvise and have a go anyway.

Objective 4:

Pippa must show what she brought with her and why. She does things properly and logically.

Objective 5:
Gloves and clean water were also brought along to avoid any risk (slight) of infection.

The next thing to understand is that only after knowing what I had filmed under the microscope, could I start to derive a rough structure to use for a script. This is an odd thing to do for a film maker because normally, if making a feature length thriller and scripting it yourself, you would write the script first and then start filming. With microscopy filming, it is not so easy. One has to know what core material (video taken at the microscope) you were able to obtain first. It would be no good putting in a section on rotifers giving birth if I had no video of that to show it.

Writing the script
A script is a both a guide for you and anyone helping you to create your film. By writing one, you can order your thoughts, plan out what you need and when, and see what you need to do for each bit of filming. Sometimes, when you actually shoot your movie, things don't always go to plan. You can adapt any changes you make back into your script in pencil so that later, when editing the film together from all the clips, you have a reminder of the change made on the day. I've lost the script I wrote for Pippa's Progress Episode 3, but I still have the script for episode 2. A sample from it appears on the next page.

The action is divided from the dialogue. Anyone speaking has their dialogue clearly separated and cantered below the speaker's name. This helps the actor/speaker identify anything they need to learn to say and helps the film maker ensure nothing is omitted during the filming. I use a scripting editor but you can write your script using a simple text editor. Just remember to format it similar to mine so it's clear for everyone to follow.

Sample of script for Pippa's Progress Episode 2:

She points at the stage.

> PIPPA (CONT'D)
> Anything you want to look at goes
> on the stage.

She picks up a slide and places it under the clips.

> PIPPA (CONT'D)
> You move the stage using this
> control. (*She does it*). It takes a
> while to get used to moving it
> because when you look down the
> microscope, it appears to move in
> the opposite direction to what you
> expect.

She clicks on the under-stage light.

> PIPPA (CONT'D)
> This light shines up through the
> specimen you look at. Most of the
> things you put on this microscope
> will be transparent. You need to
> shine the light through it to see
> anything.

She turns it off. She turns on the top light.

> PIPPA (CONT'D)
> This light, you use when the thing
> you look at is not transparent. You
> will see which one to use when we
> start looking at stuff later on.

SHE LOOKS UP AT US. She looks concerned...

[Silence]

She smiles.

> PIPPA (CONT'D)
> Maybe, it'll be more interesting if
> I show you how to use it by finding
> something to look at and how to...

Drama

No-one wants to sit down and watch a rotifer 'munching' away for half an hour or listen to 15 seconds of a droning monologue and 29 minutes and 45 seconds of whatever else is going on in your room when filming. Use close-ups (remember to exploit different levels of magnitudes when filming at your microscope), and inter-cut to other material—diagrams, easy-to-understand and interesting analogies—to liven up and enhance viewer interest.

So, a main video clip of a rotifer eating might go something like this when scripted...

Rotifer seen with cilia swirling in the water (low magnification).

> SPEAKER (V.O)
> Rotifers use their cilia for both locomotion
> and to draw food into them.

[CLOSE UP]: Cilia movement.

> SPEAKER (V.O)
> The spinning cilia create strong micro
> currents in the water.

CLOSE UP]: algae being caught up in the current with some missing the mouth and some being swept in.

[PULL BACK—LOWER MAGNIFICATION]: Whole rotifer. Foot anchored to plant of surface of glass slide.

> SPEAKER (V.O)
> The rotifer has a foot which it can use to anchor
> itself when feeding.

[CLOSE UP]: The rotifer foot.

> SPEAKER (V.O)
> The foot can be one of several designs.

[SPLIT SCREEN]—Small stills to right of main video showing different types of feet.

> SPEAKER (V.O)
> The foot may be one with toes to grasp

plants to so the animal can anchor
itself, or it may be a type which
adheres to surfaces through a substance
secreted from a cement gland.

[CUT TO]: Rotifer jaws. Close up of algae being mashed by
pounding jaws.

SOUND EFFECT: water rushing. Dull thumping noise.

SPEAKER (V.O)
The jaws of the rotifer consist of two
powerful hard pieces of material
which pound together...

[CLOSE UP]: Algae caught momentarily in the closing jaws,
mashed. The jaws open, the flattened algae is swept further
inside and ingested.

SPEAKER (V.O)
...mashing any unlucky algae
before final ingestion.

You can see from this how I would use cuts between close-ups and
whole view shots, paced and matched with the information provided by
the speaker's dialogue, to maintain a sense of action and continuity. We
are used to quickly changing viewpoints when watching videos and
films, with each new generation of people seemingly able to master an
ever faster change of pace. Quite simply: people get bored very quickly
if the film remains static.

Abbreviations used in the script are standard:

(V.O.) = Voice Over.

Recording equipment—visual recording
At the microscope itself, you can use the same equipment—camera,
attachment adaptors—as in *Chapter 8. Contemporary Digital
Photomicrography* so long as any camera used can record video as well
as stills. If you intend to show your film on the Internet, you should try
to use a camera which can record at 24 or 25 fps (frames per second) at
a definition of 1280 x 720 pixels. This will provide good resolution and

good streaming capability. If your camera can record at 1920 x 1080 pixels, better still, and you should record at this higher resolution. With this format, your film can be broadcast on television, or included in a BlueRay disc or HD DVD. You can down-scale it to 1280 x 720 for web streaming.

Alas, if your camera cannot produce these sizes then you have to go with what you have. We are all limited by our budgets with regard to equipment and resources. Lower sized videos can be scaled up using relevant software or editing packages but they will suffer from a varying degree of empty resolution and pixilation.

Macro-photographers will use similar cameras with installed macro lenses or adaptors but they have a different set of issues to resolve. It's not until you try videoing a bee taking pollen from a garden flower that you discover a slight breeze is your worst enemy. Flowers sway in the moving air—your bee along with it. Only a slight movement to and fro can take everything out of focus and back in again... to and fro!

You have to be creative and imaginative. For example, you can shoot film further away of bees flying over many flowers, and then cut to one close up in full macro. The one you cut to of course will have been positioned somewhere (picked and set up) by you earlier. You might have tied it to a garden cane to keep it from swaying, and just ensure you don't catch the cane on film so the illusion that the flower is part of the rest is maintained.

Recording equipment—sound recording
If you are going to record sound from the external environment, bees buzzing, birds singing, or streams rushing by, you need a good microphone or a separate sound recorder. Using camera microphones for recording audio for videos of macro and microscope work is not really a workable option. All the fussing around the camera focusing it means it will be recorded, and end up as unwanted off-putting noise.

You can plug external microphones into some of the digital cameras. I found the best solution though is to use an independent recording device. The one I have can be put on a tripod or a pole which is required to record someone walking around talking. The only issue then is you have to synchronise the sound recording from that device with the video footage later on to achieve a final video. This is normally done clip by clip rather than a whole film in one go. This is only necessary for recording someone speaking when they are on camera or for anything else on view where a sound is expected—a babbling stream for example—although you can sometimes just edit in a sound effect for this later. The recorder I use is a **Zoom H2N**, which records digital tracks onto a flash memory card. There are various recorders in the Zoom family and many alternatives made by other companies.

The price often reflects quality, with the more expensive recorders really designed for recording good quality music. Cheap brands are likely to introduce hiss and unwanted noise into your recording so beware of buying cheap alternatives.

Synchronising sound to video—the best way

First: the old-fashioned way

Where you need to video someone speaking and record them on a separate recording device, the old fashioned way is to use a clapper board. *If you do not record sound on your camera and only record via the independent recording device, you will need to use this method.*

You arrange your shot. You start the video recorder running. You start the sound recorder running A helper walks up to where your camera is focused—normally on the person about to speak—and he/she reads out the shot number and then snaps the board together with a sharp distinctive clip.

Hopefully, when you get to the stage of editing your film, you can then manually align the blip on the sound track (the snap of the board) with the moment in your video where the two halves of the board come together.

In a good video editing software suite, you see a visual representation of the sound wave. The sudden blip is very obvious to see enabling you to slide the sound track along the time line until it lines up with the video frame of the board halves coming together. This is true providing you are using non-linear video editing software like Adobe Premiere.

What you write on the board is said aloud and was captured on both video and sound when you shot the clip, so you can sift through all the video clips and match them up with the sound clips from the recorder. This is a bit of a frustrating process if you have a lot of clips to sync.

The modern way
Providing you record sound on your camera, onto the video, as part of the video clip, and no matter the poor quality and fumbling noises, you can automatically sync your video to the good quality sound you also recorded on the independent sound recorder. It's done magically by a software program called **DualEyes.** The software lets you select a bunch of sound only files and a bunch of video (with sound) stored on your computer and then it—the software—marries up and replaces the sound on the video clip with the correct audio only sound recorded by the independent recorder. Sometimes it fails and then you may have to do that one manually.

DualEyes was originally made and sold by Singular Software, but at the time of writing is now sold by Red Giant. It may also become assimilated into similar software called **PluralEyes**. Obtain it from the Red Giant web site here:

http://www.redgiant.com/products/

Recording equipment lights
You will only need lights when not videoing down the microscope. It is completely unnecessary to buy dedicated movie lights. Simple tungsten lights on stands are sold with 300 watt up to double 500 watt heads in most D.I.Y Hardware stores. These are relatively cheap. They do throw out a lot of heat though and may be less suitable for lighting live macro subjects like insects. You can often use aluminium foil to reflect natural sunlight onto subjects in the garden on a sunny day. LED lighting is improving all the time with costs falling and light output rising. They're cool and are becoming good alternatives to hot tungsten lights. Colour acetates can be clipped in front of tungsten lights to convert the yellowish light to daylight frequency light (blue).

Video editing
Video editing software comes in cheap and expensive bundles. Few are really free. You can only edit your film effectively if you use a non-linear editor which allows you to assemble all your clips in any order you like and shuffle them around until you're happy with the result. I cannot provide you with a whole tutorial on video editing as it would take a book longer than this one to do just that.
I can tell you the standard tips though. Follow these steps.

1.   Synchronise all your video clips with the audio ones using DualEyes or PluralEyes first. (You can't do this if you are manually synchronising).
2.   Run through the clips. Do you need them all. Get rid of any clips with errors in them. They will only take up disc space on your computer.
3.   Open your Video editing software and import all the clips. (If your film is longer than 20 minutes, it may be best to break it up into sections and work in sections first. You can stitch the sections together later).
4.   Do a rough assemble on the software timeline.
5.   Check out how that flows in principle.
6.   Rearrange if you're not happy. When you are, begin more detailed editing.

Sound effects and music
If you are going to use music in your final film, you have to be aware of copyright issues. You can't simply rip music off your CDs or your MP3 downloads and copy them or parts from them into your video editor. Well, not unless you pay a music licence fee. To do so is illegal and you can be heavily fined if you do that and use them on the Internet or on a television broadcast.

Many web sites exist which will licence you 'Royalty-free' music. Costs vary and might make that option too expensive for you. Similar sites exist where you can licence sound effects too, and the same issue with copyright and fees apply. You may not be musically gifted to create your own music tracks but it is much easier to make your own sound effects, like doors closing, crashing waves etc. Just go and record them with your audio recorder separately.

You add the music and sound effects normally once you have the existing video and sound in your film edited and when you are satisfied with that.

Before I finish this chapter, which hopefully serves as an introductory guide to any amateur microscopist who wants to make great videos at the microscope, I would like to help you (inspire you?) consider how to become a very wealthy film maker or script writer. It's simple. Write a script or make a movie based on the creatures in the microscopic world. What dinosaur study did for Jurassic Park— rotifers, desmids, stentors and the like can do for your block buster movie. Modern 3D modelling and CGI graphics have evolved enough to have rotifers the size of nuclear submarines come thundering at you on a 3D cinema screen.

Maybe shrink your heroes down into a miniature submarine in a plot which requires that and let them loose on a mission in a garden pond or stream. Toys, comics, games will all ensue from a successful film of that time and it's an untapped area.

When you make your millions, give me a mention on your TV interviews and a plug for my book and I'll be eternally grateful.

List of resources for video editors

Video Editors

Adobe Premiere (expensive but great):
http://www.adobe.com/uk/products/premiere.html

Sony Vegas Pro (less expensive and a rising star)
http://www.sonycreativesoftware.com/vegassoftware

Final Cut Pro (for Apple Mac and on a par with Adobe Premiere)
https://www.apple.com/uk/final-cut-pro/

Windows Movie Maker (*Free*. I've used it. It's ok.)
http://windows.microsoft.com/en-gb/windows-vista/getting-started-with-windows-movie-maker

Lightworks (*Free* and it seems to have enough functionality to make it a good choice if you can't afford a heavyweight editor).
http://www.lwks.com/

Chapter 13. How To Share your videos

Most people will have heard of Youtube and yes, this is a great place to share your microscopy videos with others. But people who are serious about their creations also use another web site. It's called Vimeo and the address is: *www.vimeo.com*

This is a great place where like-minded film makers of all age groups share their films with other vimeo members and non-members. There are free options for joining and paid options which allow a serious user to charge a fee for people to stream your films to their mobile devices and computers.

Whether you choose Youtube or Vimeo or both, you can share what you upload to them on Faceook and other social network sharing sites quite easily.

Or what about your very own web site? A lot of people can be daunted by the prospect of writing an HTML formatted web page but today, like never before, it's much easier using online sites that provide easy-to-use tools. One great place is WIX at: *www.wix.com*

Over 40 million people use this site to create their own web sites and it's **free.** If you use Youtube or Vimeo to host your videos, it's easy to inject your film into your Wix web page without needing to upload it again.

Formats
Streaming videos use a lot of internet bandwidth, but speeds in countries with fibre optic broadband networks into homes and schools will have no problems watching videos on-line. And the rates, speeds, and coverage improve exponentially in short time spans throughout the world. Two widely used formats for web-based video are .mp4 and .mov files. Both can be streamed and played on both Apple and Windows devices. HD (High |Definition) on the web normally means a video size of 1280 x 720 pixels but both **youtube** and **vimeo** will convert your videos to appropriate streaming formats if you made your movie in another format or larger size.

Languages
English is the most widely used International language. If you want to reach the widest audience on the Internet and your film has dialogue in a different language, it's a good ideas to create English subtitles and overlay them on your film. Most video editing packages allow text to be overlaid onto your video. When you do this, remember to put your text in what is called the 'Safe' area of the screen. This is to ensure if people watch your film on devices which shrink the screen size, or play

your movie back via streaming to an older style television screen of 720 x 576 pixels. your text is not cut off. You may also wish to convert your film and make a DVD of it to give to a friend or even sell. The normal DVD format is still 720 x 576 pixels. [Below] Outer dark grey is the full active screen size. The white area is the safe area for text. Put your text where I show you below.

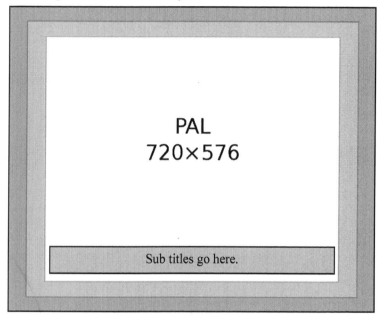

It's best to use a simple and common format text style for your sub titles: Times New Roman is a commonly used and an easy to read font. Fancy text can be difficult to read or may make your video unprofessional looking. Videos use codecs to encode and compress visual information for your film The best one to use at the time of writing is one called H.264

Changing Technologies
It's important to realise new formats, codecs, film resolutions, and methods change all the time. The film maker can be faced with a puzzling array of different systems, tools, and methods. As time passes, check forums and web sites for the best emerging ways of making and presenting your microscopy films and images.

Chapter 14. Emerging Ideas & Technology

Humankind's ingenuity is truly staggering. Whenever the pragmatists in our midst, or experts in any field become entrenched in a traditional and unchanging philosophy or perspective—some bright spark comes along and turns everything about that field on its head. So, what does that mean for the future of amateur optical microscopy?

Plastic microscopes

A decade or so ago, the open university microscope was in existence. It was lightweight, could be sent out to students all over the UK relatively cheaply and it didn't break easily.

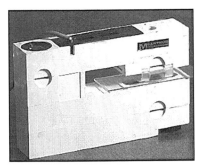

The design was based on the McArthur Field Microscope, a wonderful optical microscope.

What many people, especially in the west, do not realise is that in many so called 'third world' countries the prevention and early detection of common diseases and parasites could save millions of people from terrible suffering and death.

The problem is microscopes cost money. They cost money to make and money to ship. If only we had a microscope that was cheap to make, cheap to send, and easy to assemble, we could remove much of that suffering.

Paper microscopes?

Paper is lightweight, inexpensive, and thus would be an ideal material to make microscopes. And they would be cheap to post too. Impossible you say? Well, it seems not. How about a paper microscope that an end user pops out of a thin card and folds like origami to form a 1600x magnifying optical microscope. More... imagine it can be made for less than one dollar. Well, an intelligent and imaginative person has designed one and they are now being made and tested. *Enter the Foldscope!*

It's is an origami-based print-and-fold optical microscope that can be assembled from a flat sheet of paper. Although it costs less than a dollar in parts, it can provide over 1,000X magnification with sub-micron resolution (800nm), weighs less than two pennies (8.8 g), is small enough to fit in a pocket (70 × 20 × 2 mm), requires no external power, and can survive being dropped from a 3-story building or stepped on by a person. Its minimalistic, scalable design is inherently application-specific instead of general-purpose—gearing towards applications in global health, field based citizen science and K12-science education.

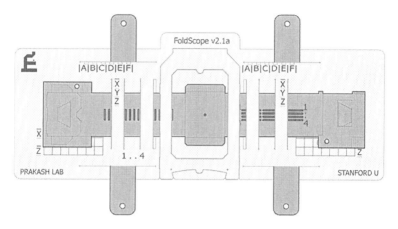

Visit the website for more details: *http://www.foldscope.com/*

3D Printers

A new technology is just now becoming affordable for the home consumer: 3D printers. Basically they use similar ink-jet techniques to deposit tiny droplets of molten plastic to create a 3D form. Car parts, aeroplane parts, and many other items can now be printed out and made by them. What about a 3D printed microscope? Done!

Well almost. You have to source the glass lenses yourself. But what is clever, is the idea a computer program or a design instruction set can be sent or downloaded through the internet and exploited on a device which manufactures something.

We are on the verge of 3D digitally printed skin, for example, or any range of other things by clever use of new 'printable' materials. **Download your 3D printable microscope from:** (see over)
http://www.thingiverse.com/thing:77450

How about something which scans a specimen slide of human tissue which then converts that into a 3D digital printer instruction set such that the receiving printer deposits granular material onto a glass slide to replicate it? The resulting slide may contain an intrinsically transparent plastic component but which can be in such a way that the structure, its varying density and transparency, can exactly mimic the original. But it will not need fixing, staining, or preserving. And... it (the design) could be sent to thousand of end users. One slide replicated and studied by many.

Submarine Microscope
Instead of bringing things from ponds and streams to study on our microscopes, what about a mini Radio Controlled submarine to drop into that village / garden pond equipped with a suitable microscope and light to go study those things in their habitat? Could we design one, right here? We can at least learn a bit about what the requirements might be. My first thought would be: is there anything already made we can adapt or do we have to start from new?
How about this..?

All the spec is there. We could possibly put that camera on a small optical microscope system and use it to send us video from what it finds. This little marvel doesn't come cheap. The assembled version costs approx. £400.00 from here: *http://www.marionvillemodels.com/rc -shop/thunder-tiger-neptune-rc-submarine.html* but there are kit versions you assemble yourself. This is certainly not some plastic toy. Don't believe me? Download the tech spec and assembly sheets here:

http://mms.tiger.tw/upload/upload_file/Boat/5220%20JJ6072V4.pdf

High intensity mini-LED-floodlight come as accessories. Now surely some bright spark out there can put a relatively simple optical microscope front on that camera (what about the lens system on the Foldscope. Could it be cut out and fitted to the sub?

Application of Microscopy in art & engineering
Want to design a military helmet, a lunar walker, irrigate a desert, or maybe none of those, but beautiful art pieces instead? Nature has been working flat out for 3.5 billion years testing out various designs. Most of them never worked out, but the ones that did do well are all around us now. And some of the best are there for all to see down the end of a microscope or in the macro world of bugs and micro plants.

Look at cross section stems of bamboo, reeds or any other plant which rises up in a single stem unsupported. Can that design be used to build mobile phone masts, sky-rise building, sailing masts? Do you see my point?

Want to irrigate a desert or a dry pasture with the minimum length of waterway? Plant and tree leaves do it all the time. Look at their veins. The body of a harvester spider hangs suspended on the frailest, springiest legs. The next time you're under your sink or looking into a dark cupboard in your toilet, if you have one, chances are those spindly, furtive creature in there are Harvesters.

You won't read what I am about to tell you in any books. Hardly anyone looks at small things properly except us amateur microscopists.. oh... and a few professional microscopists too. Harvester spiders can go *invisible!* Yup... invisible.

I'm not going to tell you how they do that but I watch them do it all the time. It is one of their defence mechanisms. If you want to see them do it—very, very gently move your finger up slowly and let the surface of it just barely touch one of its legs near the base. It might just move away at first. Try it again. If you are lucky, it won't. Watch how it goes invisible. Do you think it has something to do with those wispy legs and the way the body is so low slung and balanced between them?

Art

Inspirations galore here for the artists of all styles: abstract, real, or surreal. I am amazed by how many people who love to draw, paint, or make snappy, cool tee-shirts with cheeky words and pictures emblazoned across them, rarely use the fantastic images available in the very small world. I think I'll design a tee-shirt...

BIG BROTHER
SOMETHING

IS WATCHING YOU

There we lave it then. But a young guy or girl with an eye for what's happening on the street could certainly make a satirical comment better than mine.

And that's it for this book. I hope you are inspired and suitably informed of the hobby of amateur microscopy and if you fancy dropping me a line anytime, I'm always happy to hear from you. Write to me at::

molsmith@fastmail.fm

Or visit our web site and magazine at:

www.microscopy-uk.org.uk

And...

www.micscape.net

"Happy Scoping!"

Mol Smith

Come and learn with PIPPA at: **www.pippasprogress.net**

APPENDICES

Appendix 1: References & Resources

Chapter 1: What you need to start

UK Recommended company– Brunel Microscopes	www.brunelmicroscopes.co.uk
UK Buying accessories and glass slide	www.brunelmicroscopes.co.uk
Books: Introduction to Microscopy	Adventures With A Microscope ASIN: B0006D6PSM Price: £10.00 approx.
Books: Introduction to Microscopy	The World Of The Microscope Paperback: 48 pages Publisher: Usborne Books Language: English ISBN-10: 079451524X ISBN-13: 978-0794515249 Price: £6.00 approx

Chapter 2: How to practice Microscopy

POND LIFE—WARD AND WHIPPLE Comprehensive Illustrations for identifying microscopic life in water	www.microscopy-uk.org.uk/mag/ artnov13/ms-whipple.html

Great web sites:

1. Established in 1995 International web site

   supporting amateur microscopy — www.microscopy-uk.org.uk

2. Micscape Magazine & online library — www.micscape.net

3. Microscopy clubs & forums — www.micscape.net
   Also see Appendix 2 for clubs!

4. Cells Alive. Great for children and adults — www.cellsalive.com

5. Molecular Expressions. Comprehensive — www.micro.magnet.fsu.edu

6. Micropolitan Museum. Kids love it. — www.microscopy-uk.org.uk/
   micropolitan/x_index.html

Chapter 3: How to use your Microscope

Pippa's Progress—Free Video Tutorials	www.pippasprogress.net
Primer on using a Microscope	www.microscopy-uk.org.uk/ primer/

Chapter 4: First Projects

Paraffin Wax Processing DVD	www.brunelmicroscopes.co.uk/ microtomy.html
Micrographia Online by Robert Hooke	www.gutenberg.org/ files/15491/15491-h/15491-h.htm

Appendix 1: References & Resources *Continued*

Chapter 5: Slide Mounting

Frutose Sugar. Where to buy   Health food shops

Glass slides & Cover slips (UK)   http://www.brunelmicroscopes.co.uk/
                                               coverslips.html

SAFE MICROSCOPIC TECHNIQUES FOR   http://www.amazon.co.uk/MICROSCOPIC-
AMATEURS Slide Mounting [Paperback]   TECHNIQUES-AMATEURS-Slide-
ISBN-10: 1499746512   Mounting/dp/1499746512
ISBN-13: 978-1499746518
Paperback: 102 pages

Chapter 6: Pond Life

Nil

Chapter 7: Microscopy History

Robert Hooke-father of modern science   www.microscopy-uk.org.uk/mag/artmar00/
                                              hooke1.html

Nineteenth Century British Microscopyand   www.microscopy-uk.org.uk/mag/artmay08/
Natural History. y Richard L. Howey,   rh-british5.html
Wyoming, USA

Chapter 8: Contemporary Digital Photomicrography

Raynox DCR-250 Macro Attachment   Search internet for best prices—approx
Fits most cameras. No macro lens requited! £40.00

Focus (image) stacking software.   www.hadleyweb.pwp.blueyonder.co.uk
Combine CZ

Microscope adaptor for cameras without   www.microscopyimaging.co.uk/
removable lenses—Linkarm   brunelunilinkadpaters.html

Microscope adaptor for cameras without   www.microscopyimaging.co.uk/
threaded lenses—UNIlink   brunelunilinkadpaters.html

Photostitching software—a list of   *http://en.wikipedia.org/wiki/*
                                              comparison_of_photo_stitching_software

Zoomify (Zooming large images)   http://www.zoomify.com/

Chapter 9 & 10:& 11
3D Microscopy

Prints of your lenticular images.   www.3dz.co.uk/

An online list of resources,   http://lenticular3d.com/

Software:S tereoTracer Software &   http://triaxes.com/
3DMaster Kit Software

Appendix 1: References & Resources *Continued*

Alternative programs and software for stereo, 3D, and lenticular imager creation.   www.stereo3d.com/3dhome.htm

Free 3D software—recommended!   www.stereo.jpn.org/eng/

An online 3D virtual microscope.   www.microscopy-uk.org.uk/3d-microscope/

Posing and animation of models:

Poser (Posing & animation)   http://poser.smithmicro.com/

DazStudio (Posing & animation)   https://www.daz3d.com/get_studio

Turbosquid (3D models)   http://www.turbosquid.com/

Model creation software. Comprehensive list   http://3d.about.com/od/A-Guide-To-3D-Software/tp/Full-3D-Suites.htm

Microscopy-related 3D models in action on my web site: about them   www.microscopy-uk.org.uk/mag/artjan10/moll.html

3D models I've used   www.microscopy-uk.org.uk/3dmodels-index.html

Chapter 12. Making Microscopy Movies

Video Audio Equipment —Zoom Sound recorder.   google ZOOM H2N or just ZOOM Recorders

Video / Audio Syncing. Dual eyes / Plural Eyes   www.redgiant.com/products/

Video Sharing Sites—youtube   www.youtube.com/

Video Sharing Sites—vimeo (recommended).   www.vimeo.com/

Video Editors:

Adobe Premiere (expensive but great):   www.adobe.com/uk/products/premiere.html

Sony Vegas Pro (less expensive rising star)   www.sonycreativesoftware.com/vegassoftware

Final Cut Pro (for Apple Mac and on a par with Adobe Premiere)   www.apple.com/uk/final-cut-pro/

Windows Movie Maker (*Free*. I've used it. It's ok.)   http://windows.microsoft.com/en-gb/windows-vista/getting-started-with-windows-movie-maker

Lightworks (*Free* and it seems to have enough functionality to make it a good choice if you can't afford a heavyweight editor).   http://www.lwks.com/

Appendix 1: References & Resources *Continued*

Misc. Resources

Foldscope—the origami Microscope	http://www.foldscope.com/
Print a microscope with 3D Printer	http://www.thingiverse.com/thing:77450
RC Model Submarine—specification	http://mms.tiger.tw/upload/upload_file/ Boat/5220%20JJ6072V4.pdf
RC Model Submarine—purchase	http://www.marionvillemodels.com/rc-shop/ thunder-tiger-neptune-rc-submarine.html
My online 2D Microscope (again?)	www.microscopy-uk.org.uk/2D-microscope/

Appendix 2: Clubs & Forums

Macro Forums

Macro photographers (Forum)	www.photomacrography.net
Macro photographers (Forum)	www.thephotoforum.com/forum/macro-photography/
Macro photographers (Forum)	http://macronatureforum.com/forum/

Microscopy Forums

Quekett (for members and non-club people)   www.quekett.net/forum/index.php

Yahoo Groups:

Amateur Microscopy	https://groups.yahoo.com/neo/groups/amateur_microscopy/info
Diatom Forum	https://groups.yahoo.com/neo/groups/diatom_forum/info
Microscopes	https://groups.yahoo.com/neo/groups/Microscope/info
Wild Microscope Users	https://groups.yahoo.com/neo/groups/Wild_M20/info
Italian Speaking Forum	https://groups.yahoo.com/neo/groups/microcosmo_italia/info
CombineZ Focus stacking forum	https://groups.yahoo.com/neo/groups/combinez/info
Germany based Forum	http://www.mikroskopie-treff.de/
France based Forum	http://www.lenaturaliste.net/forum/index.php

Clubs & Societies Web Sites, & Misc.

Royal Microscopy Society Journal	http://www.rms.org.uk/
Quekett Microscopical Club Journal	http://www.quekett.org/
Light Microscopy links and articles	http://www.microimaging.ca/
Little Imp Publications public domain books on CD. (Hosted by Savona Books)	http://www.savonabooks.free-online.co.uk/
Mikroskop Museum German website	http://www.mikroskop-museum.de/
Molecular Expressions Web site	http://micro.magnet.fsu.edu/
Dennis Kunkel's SEM images	http://www.denniskunkel.com/

141

Appendix 2: Clubs & Forums

Diatoms Ireland — http://www.diatomsireland.com/

Leitz museum — http://www.leitzmuseum.org/

Microbe hunter a free monthly microscopy enthusiast magazine — http://www.microbehunter.com/

Micscape Magazine—bimonthly — www.micscape.net

Microscopies online magazine and Forum for French speakers — http://www.microscopies.com/

Micrographia articles, projects for the enthusiast — http://www.micrographia.com/

Microscopy Today journal with online archive — http://www.microscopy-today.com/

Modern Microscopy online journal by McCrone Group — http://www.modernmicroscopy.com/

Fun Science Gallery projects, articles on optics — http://www.funsci.com/

G. Couger's microscopy links, extensive links — http://www.couger.com/microscope/links/gclinks.html

A Cabinet of Curiosities —a resource on Victorian microscope slides by Howard Lynk — http://www.victorianmicroscopeslides.com/

Science Projects by Ely Silk — http://www.viewsfromscience.com/

Historical makers of microscopes and microscope slides Brian Stevenson's extensive resource on Victorian slide makers — http://microscopist.net/

Made in the USA
Lexington, KY
24 November 2014